HashiCorp Packer in Production

Efficiently manage sets of images for your digital
transformation or cloud adoption journey

John Boero

BIRMINGHAM—MUMBAI

HashiCorp Packer in Production

Group Product Manager: Preet Ahuja

Publishing Product Manager: Preet Ahuja

Senior Editor: Arun Nadar

Technical Editor: Rajat Sharma

Copy Editor: Safis Editing

Project Coordinator: Ashwin Kharwa

Proofreader: Safis Editing

Indexer: Hemangini Bari

Production Designer: Shankar Kalbhor

Marketing Coordinator: Rohan Dobhal

First published: June 2023

Production reference: 1140623

Published by Packt Publishing Ltd.
Livery Place
35 Livery Street
Birmingham
B3 2PB, UK.

ISBN 978-1-80324-685-7

www.packtpub.com

To my parents, Joe and Rose Boero, who have tolerated my shenanigans for all these years.

Foreword

Today, HashiCorp is known for building a broad portfolio of products used for infrastructure automation. What most don't know is Packer was the first tool we created after starting the company in 2012. The timing at which Packer was released made it rather controversial, as that was the heyday of configuration management and it was common practice for organizations to have long-running infrastructure that was patched and upgraded over time.

Early in the history of HashiCorp, we published the Tao of HashiCorp, which was our set of strongly held principles for managing the infrastructure around which our products were built. One of those key principles is immutability, which we felt was a critical way to manage complexity and risk in large-scale infrastructure. One of the key challenges with mutable approaches to infrastructure is that you have no discrete versions and can't reason about what portion of the fleet is running version 1 or version 2. Instead, you have a continuous spectrum of versions, with partial failures and drift, forcing operators to reason about version 1.5 or version 1.87.

Immutability provides a path out of this madness. By moving our configuration management, patch management, and machine hardening into pipelines that are preproduction, we get golden images that we can version. That allows us to have a sensible inventory of our production environments while eliminating an entire class of complexity and risk that comes with partial failures. Of course, that creates an entirely new set of challenges. Now, we need to be good at building and rebuilding images frequently and across many different platforms. This was the problem Packer was built to solve, by giving us a common workflow to define and build machine images across many different platforms.

Packer has since become an enormously successful product that is used by thousands of organizations as a key part of how they build and manage production infrastructure. Beyond direct users, Packer has influenced cloud providers as well, and services such as Azure Image Builder are built using Packer.

Over the years, Packer has evolved to solve new challenges. Docker and OCI images came to be several years after Packer was released, and it has evolved to support those formats as well. The irony is that Docker helped bring immutable infrastructure into popular practice more broadly. Today, HashiCorp continues to invest in Packer both with the tool itself and through HashiCorp Cloud Platform, or HCP, with the HCP Packer service. HCP Packer provides an image registry with metadata about images, versions, and artifacts that can be queried by downstream automation tools such as Terraform and Ansible. This enables an end-to-end automation pipeline from the image building to the tools that deploy and manage the production infrastructure.

As an industry, we continue to evolve our approach and best practices around infrastructure management. I feel strongly that immutability has a key role to play in managing at scale, and that Packer plays a critical role in the toolchain. I hope this book is a useful resource for readers to learn more about Packer and how it can be used.

Armon Dadgar

Co-Founder and CTO of HashiCorp

Contributors

About the author

John Boero is a solutions engineer with HashiCorp in London covering partners across the EMEA region. He has worked for Red Hat, Puppet, and others. He has held roles as a developer, DBA, consultant, and architect all around the globe, whether San Francisco, Canary Wharf, or Saudi Arabia. He is an infrastructure minimalist and loves to push and bend projects beyond their standard use cases. He is active in the Fedora project and the wider open source community. Of all of the HashiCorp products, he chose Packer to write a book about because of its simple yet powerful functionality and its recent rewrite to make use of HCL2.

I want to thank the people who have supported me during this long writing process, including the Packt team and my friends and family who put up with the long nights and weekends I spent writing sample code.

About the reviewers

Clara McKenzie is currently a team lead for infrastructure at Rebellion Defense with a long career in software, specializing in networking and embedded platforms. Born in San Francisco, she received a BA in mathematics from Reed College in Portland, Oregon. Her work adventures include the NFS/RPC group at Sun Microsystems, the NetWareForMac team at Novell, Core Software at Ascend Communications, the Gracenote SDK team, Planet Labs Dove Satellites, and supporting Terraform at HashiCorp. She goes by the handle @cemckenzie on GitHub.

Rickard von Essen works as a continuous delivery and cloud consultant at Diabol. There, he helps companies deliver faster, improve continuously, and worry less. He was one of the early maintainers of Packer and contributed to the Parallels plugin. In his spare time, he flies his FPV drones, repairs retro computers, and contributes to numerous FOSS projects. He has been tinkering with Linux and BSD since the late 1990s and has been hacking since the Amiga era. When not fiddling with computers, he spends his time with his wife and three children in Stockholm, Sweden. Rickard received an MSc in computer science and engineering from Linköping University.

Table of Contents

3

Configuring Builders and Sources   23

4

The Power of Provisioners   41

5

Logging and Troubleshooting   53

Part 2: Managing Large Environments

6

7

8

Part 3: Advanced Customized Packer

12

Developing Packer Plugins 145

Index 157

Other Books You May Enjoy 164

Preface

As a frequent blogger of HashiCorp products, it's an honor to present this more thorough exploration of the open source Packer tool. Packer is one of HashiCorp's simplest and most elegant tools with one simple task. It builds images and artifacts in a consistent way for different environments and cloud projects. If you're thinking of carrying out a lift-and-shift migration, there are a series of questions you may have:

- What operating system will you boot with?
- How will you apply updates and test patches across your development lifecycle?
- How can you ensure your on-premises images match the ones in your global cloud environments?
- What if you have multiple clouds to keep track of?
- How can you build matching container images for serverless environments or orchestration platforms?

These are the questions that Packer solves for most users. On the journey of writing this book, I went even deeper to try and show how Packer can do even more than it was built for. I wanted to answer the questions that people don't think to ask about Packer. In addition to cloud environments, we'll explore everything from the high-level basics of local environments to complex cloud image libraries. We'll even push Packer further and experiment with some interesting use cases Packer hasn't necessarily been positioned for. The questions you'll be asking afterward may surprise you:

- Are your cloud applications prepared for ARM or RISC-V architectures to cut your compute costs in half?
- Can you run tests and compliance profiles as part of your automation pipelines?
- Can you build and test mobile apps along with your cloud images?
- Can you build and test your IoT devices and microcontrollers along with a cloud backend?
- Can Packer save you time on slow Terraform tasks?

The answers to these questions are surprising, and it turns out Packer can do every one of these things. It can replicate your local VMware environments in the cloud for lift-and-shift workloads. It can update your container image registry at the same time as it builds a virtual image or even a bare-metal environment for storage or network booting. This book and the corresponding GitHub repo show you how.

This book starts off with a foundation of manual builds, working our way up to build automation. Build automation can help speed up your development experience, so you may want to go back over the book in reverse using automation.

I would like to thank a few members of the Packer engineering team who have occasionally tested, provided feedback, or helped guide some of the sample code for this book, namely Megan Marsh, Wilken Rivera, and Lucas Bajolet.

Any author commissions for this title will be donated to the Raspberry Pi Foundation to encourage kids to learn how to code.

Who this book is for

This book is for DevOps engineers, cloud computing customers, and even students or people who want to learn about computing platforms and infrastructure as code. Thanks to the platform-agnostic way HashiCorp projects are built with Golang, every example or use case within this book can be run on anything from a multi-million dollar mainframe to a basic cloud instance, or even a 10-dollar Raspberry Pi device. We will also cover automation, which can be explored with free online accounts.

This book is also for veteran Packer users who have developed templates with the old JSON format. Packer version 1.7 introduced two new format options for templates. A lot of existing example code and documentation are now supported as legacy. This book helps with the transition to HCL2 or PKR.JSON for new template development.

We will also dive a bit deeper and cover the basics of writing Packer plugins in Golang for those who encounter a niche or new use case that they would like added to Packer's features.

What this book covers

Chapter 1, *Packer Fundamentals*, introduces Packer to those who have never used it before. This chapter will look at the use cases, architecture, and history of the project, as well as covering the basics of a Packer build process.

Chapter 2, *Creating Your First Template*, covers a traditional Hello World equivalent code example for Packer. This establishes the basic context for starting any Packer template.

Chapter 3, *Configuring Builders and Sources*, explores how to invoke one or more instances of a builder using sources in HCL2 syntax.

Chapter 4, *The Power of Provisioners*, describes how once an environment has been defined in a source, provisioners load or run customization to set up your desired applications and artifacts. The name *provisioner* comes from ships loading what they need for a voyage, a concept common to HashiCorp's Vagrant, Packer, and Terraform products.

Chapter 5, Logging and Troubleshooting, highlights how, during the development process, it's inevitable that troubleshooting will need to be done when builds go wrong. This isn't always clear when doing multiple complex builds.

Chapter 6, Working with Builders, takes a deeper dive into the different types of builders available, from the local builders of VMware, VirtualBox, and QEMU to cloud builders and Docker containers. We will also explore alternative architectures with QEMU , which is able to emulate a wide variety of architectures.

Chapter 7, Building an Image Hierarchy, establishes how as images share things in common or extend other images, it becomes important to implement an image hierarchy strategy. We cover extended system images and container images in a logical way.

Chapter 8, Scaling Large Builds, shows how when large or complex builds or multi-cloud builds become slow, it's time to optimize parallel builds. We will discuss parallel builds and make sure you don't have to wait longer than necessary for a build result.

Chapter 9, Managing the Image Lifecycle, covers what to do when images become old and obsolete or need urgent patching. It's important to set up a lifecycle strategy.

Chapter 10, Using HCP Packer, introduces HCP Packer, a SaaS platform from HashiCorp that allows automatic metadata and lifecycle management for the cloud. Here, we cover the basics of HCP's free tier and, optionally, the paid tier.

Chapter 11, Automating Packer Builds, applies automation to everything learned about in the previous chapters. Packer runs can run via automated pipelines using common VCS offerings such as GitHub Actions or GitLab CI.

Chapter 12, Developing Packer Plugins, shows what to do if you come across a potential use case that Packer doesn't cover. You may be able to extend the project with a plugin. This is an advanced topic that requires some understanding of Go.

To get the most out of this book

It will help to have a basic understanding of the HashiCorp Configuration Language, known as HCL2. JSON is also offered as an alternate option to some HCL2 Packer uses. Packer and most HashiCorp projects are written in Golang, also known as Go, but this will only be necessary if encountering a bug in Packer or extending Packer by writing a plugin.

The Packer binary is built for most common platforms and architectures, including Linux and Windows, and the most common architectures, such as x86_64 and ARM/AArch64. Since Packer is statically linked, it relies on no external libraries and the underlying operating system shouldn't affect the user experience unless using external plugins that aren't built for your target platform.

Software/hardware covered in the book	Operating system requirements
HCL2	Platform agnostic
JSON	Platform agnostic
Golang	Platform agnostic

If running Packer in production, make sure to install only official releases signed with the HashiCorp engineering signing keys. The code for this book requires at least Packer version 1.7. Future releases may change behavior, so please verify the newest Packer release that works for you. For development or following along with this book, you may build Packer from the source or use a binary directly from the HashiCorp releases page: `https://releases.hashicorp.com/packer/`.

If you are using the digital version of this book, we advise you to type the code yourself or access the code from the book's GitHub repository (a link is available in the next section). Doing so will help you avoid any potential errors related to the copying and pasting of code.

Every sample of code for this book can be considered platform agnostic. It can be run on Windows or Linux, or even Raspberry Pi. Adjust the code as necessary to suit your platform. Packer is designed to run on just about anything and remote builds in the cloud will not require large amounts of resources locally. Local builds or virtual machines may require local storage and RAM.

Download the example code files

You can download the example code files for this book from GitHub at `https://github.com/PacktPublishing/HashiCorp-Packer-in-Production`. If there's an update to the code, it will be updated in the GitHub repository, so please check back as changes are necessary. Also, feedback and pull requests are welcome!

We also have other code bundles from our rich catalog of books and videos available at `https://github.com/PacktPublishing/`. Check them out!

Download the color images

We also provide a PDF file that has color images of the screenshots and diagrams used in this book. You can download it here: `https://packt.link/GJ9i3`.

Conventions used

There are a number of text conventions used throughout this book.

`Code in text`: Indicates code words in text, database table names, folder names, filenames, file extensions, pathnames, dummy URLs, user input, and Twitter handles. Here is an example: "Troubleshooting errors may depend on the cloud provider, so it's a good idea to use `PACKER_LOG=1` as you develop these templates. "

A block of code is set as follows:

```
post-processor "compress" {
  output = "{{.BuildName}}.gz"
  compression_level = 7
}
```

Any command-line input or output is written as follows:

```
$ curl -X POST --data "{\"token\": \"$GITHUB_TOKEN\"}" \
  http://yourvault:8200/v1/auth/github/login \
  | jq -r .auth.client_token
```

Bold: Indicates a new term, an important word, or words that you see onscreen. For instance, words in menus or dialog boxes appear in **bold**. Here is an example: "Remember you need administrative rights to your repo to access the **Actions** menu."

> **Tips or important notes**
> Appear like this.

Get in touch

Feedback from our readers is always welcome.

General feedback: If you have questions about any aspect of this book, email us at customercare@packtpub.com and mention the book title in the subject of your message.

Errata: Although we have taken every care to ensure the accuracy of our content, mistakes do happen. If you have found a mistake in this book, we would be grateful if you would report this to us. Please visit www.packtpub.com/support/errata and fill in the form.

Piracy: If you come across any illegal copies of our works in any form on the internet, we would be grateful if you would provide us with the location address or website name. Please contact us at copyright@packt.com with a link to the material.

If you are interested in becoming an author: If there is a topic that you have expertise in and you are interested in either writing or contributing to a book, please visit authors.packtpub.com.

Share Your Thoughts

Once you've read *HashiCorp Packer in Production*, we'd love to hear your thoughts! Scan the QR code below to go straight to the Amazon review page for this book and share your feedback.

https://packt.link/r/1803246855

Your review is important to us and the tech community and will help us make sure we're delivering excellent quality content.

Download a free PDF copy of this book

Thanks for purchasing this book!

Do you like to read on the go but are unable to carry your print books everywhere?

Is your eBook purchase not compatible with the device of your choice?

Don't worry, now with every Packt book you get a DRM-free PDF version of that book at no cost.

Read anywhere, any place, on any device. Search, copy, and paste code from your favorite technical books directly into your application.

The perks don't stop there, you can get exclusive access to discounts, newsletters, and great free content in your inbox daily

Follow these simple steps to get the benefits:

1. Scan the QR code or visit the link below

https://packt.link/free-ebook/9781803246857

2. Submit your proof of purchase
3. That's it! We'll send your free PDF and other benefits to your email directly

Part 1: Packer's Beginnings

This part offers a basic introduction to Packer, including the build process and what code options are available. Early versions of Packer used only JSON, but newer releases have added HCL2 as the new standard in Packer templates. In these chapters, we will lay the foundation for beginner-to-intermediate users of Packer.

This part has the following chapters:

- *Chapter 1, Packer Fundamentals*
- *Chapter 2, Creating Your First Template*
- *Chapter 3, Configuring Builders and Sources*
- *Chapter 4, The Power of Provisioners*
- *Chapter 5, Logging and Troubleshooting*

1

Packer Fundamentals

Packer is a free and open source extensible software tool that takes your desired OS and container configurations and builds them simultaneously for the easy testing and management of complex system and application images and artifacts. If you ever find yourself in an environment where multiple custom system disks or cloud AMIs must be consistently maintained and adjusted to boot VMs or run containers, then Packer is here to simplify your life as you automate configuration through code.

This chapter is a very high-level introduction for those unfamiliar with Packer. It explains how Packer is not a service but a tool that can be manually run or inserted into an automation pipeline. It also describes how Packer can supplement Terraform to dramatically simplify anything from complex hybrid or multi-cloud deployments to on-premises private cloud or even local VMs on a development machine.

In this chapter, we will cover the following topics:

- *Packer architecture*, which describes how the Packer binary is distributed and developed and how Packer works with templates, builders, and provisioners at a high level

- *History of Packer*, which is important to understand why Packer was needed in the first place and what business problems it solves

- *Who uses Packer?*, which lists what types of users Packer has today, including everything from small academic labs to large-scale enterprise organizations and software vendors

- *Alternatives to Packer*, which is a section that describes industry alternatives and other tools that offer image management and how they compare to Packer at a high level

- *Installing Packer*, which covers how easy it is to install Packer on most environments, whether servers, cloud instances, or local laptops

- *HCL versus JSON*, which is a very high-level description of JSON and **HashiCorp Configuration Language** (**HCL**) and how Packer supports either standard for templates

Technical requirements

For this chapter, you should have a basic understanding of JSON and HCL2 domain specific languages. You won't need to try any sample code for this chapter but if you choose to follow along you may use any device that supports running the Packer binary. This includes any laptop or small device running Linux, macOS, or Windows. As we focus a lot of examples on Linux, it may be useful to run Linux or use a cloud resource running Linux at minimum.

Packer architecture

Packer itself is a fairly simple binary written in Go. It supports plugins for various inputs and outputs. The plugins that translate your configuration and scripts into artifact outputs are called *builders*. Common builders include common hypervisors such as VMware, QEMU, VirtualBox, AWS, GCP, and Microsoft Azure. Builders also include multiple container image formats, including LXC, LXD, Docker, and Podman. Many plugins have been contributed by the community and we will cover how you can write your own in a future chapter.

The bit of code you write to tell Packer what to do is called a *template*. Early versions of Packer expected your template to be written in **JavaScript Object Notation** (**JSON**). As of Packer version 1.7.0, both JSON and **HashiCorp Configuration Language version 2** (**HCL2**) are supported, with the latter being preferred. We will cover both formats and how you can migrate a JSON template into an HCL2 template shortly.

Provisioners are tasks or resources that should be applied to your image before packaging. By default, each builder in your template takes each provisioner. Take an example where you want to build a system image with your application across AWS, Azure, and GCP. All you need to do is define your list of builders for AWS, Azure, and GCP and include a single provisioner that uploads your application.

A build job is what runs the Packer build command with your template. Normally, this forks a parallel process for every builder you specify in your template. A build can happen simultaneously across VMware, AWS, Azure, GCP, or other builders while Packer tracks the results and reports any errors. When all builders finish or end in an error, the job is done and the Packer process terminates. Optionally, Packer may compress output images before terminating, to save space.

History of Packer

The origins of Packer can be found in HashiCorp's Vagrant product. Vagrant was originally a Ruby project to select from a set of standard OS images, boot one or more VMs, and automatically configure them once booted. Vagrant allowed for rapid environments for development with an extensible framework to support multiple virtualization platforms, such as VirtualBox, VMware, and QEMU.

When managing multiple environments for multiple teams, one needs to strike a balance of build time versus runtime. Provisioning resources is quick and easy when everything comes in a pre-built package or artifact, but purpose-built artifacts for every use case take up quite a bit of storage. What resources will be common across an organization and which might be deployed in different ways when they are consumed? Building multiple gold images for Vagrant or cloud environments becomes a challenge at scale. Packer was built to simplify this and it works very efficiently. It can be run simply on your own computer or it can be inserted into automation jobs and pipelines. We will cover all of these use cases in this book and show you how easily Packer can simplify your image maintenance both locally and in the cloud. A team that needs identical images built across multiple regions, multiple clouds, and possibly even local infrastructure may require complex image management. Each region within each cloud may need multiple versions of an image to be maintained, based on the OS, applications deployed, and custom configuration. Keeping all environments consistent often creates exponential complexity. Imagine each line in this diagram represents a combination that requires an image to be built and maintained:

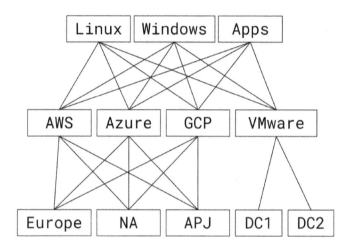

Figure 1.1 – Managing multiple applications across multiple environments can be complex

Many people will attempt to manage complex environments like this one using purely provisioning tools such as Vagrant and Terraform, which can actually result in more complexity in the end. A minor change to a Terraform provisioner can result in an entire environment being destroyed and rebuilt. It's important to start with a good image strategy before provisioning to simplify things at runtime. Often, a single Packer template can be used to satisfy all of the preceding combinations.

Packer was also the first HashiCorp project written purely in Go, also known as Golang, the modern programming language created by Google. Go is an optimized compiled language that generates simple statically linked binaries using a community of open source projects. A lot of management tools like Packer tend to be written in a scripting language such as Python or Ruby so that they can be easily ported and customized. Even Vagrant was initially written in Ruby. Scripting languages such as Ruby tend to not perform as well as precompiled Go binaries. Scripting languages are also prone to dependency deprecation and complexity. If you download a Packer binary, everything you need to run is self-contained. You won't run into an issue where an old OS version of glibc or Python prevents the binary from running. You also won't have memory leaks or buffer vulnerabilities as Go manages its own memory via garbage collection. Golang has since been the language of choice for HashiCorp projects, including Vagrant, which was rewritten in Golang for consistency. If you don't know how to write Go, there is no need to worry. You won't need to write Go to use Packer unless you want to write a plugin or add a feature. We will cover how to do this in *Chapter 12, Developing Packer Plugins*.

You can also find books on Go from Packt here: `https://www.packtpub.com/gb/tech/go?released=Available&language=Go`.

Who uses Packer?

Packer is a purely open source tool for HashiCorp but that doesn't mean that enterprise customers don't use it. Packer is used to build images in private networks and public clouds around the world, covering many industries from investment banking to universities and students. Individuals and small teams often use Packer to maintain a set of disposable system images for mixed estates, including Mac, Windows, Linux, and serverless cloud applications. Large teams and organizations may use automation or continuous deployment pipelines for Packer to rebuild a set of images when certain events or edits occur. The beauty of Packer is it behaves the same whether it is running in a multi-cloud Fortune 500 firm or running on a laptop in a coffee shop. You can even run Packer on a low-power commodity ARM device such as Raspberry Pi. The difference between a coffee shop laptop and an enterprise deployment really comes down to security and best practices, which we'll cover in *Chapters 6-8*.

The open source community has a great variety of sample templates, so you usually don't need to start one from scratch. Search the Packer documentation page for samples, as well as GitHub. Unlike Vagrant, which has a public registry of source images, Packer requires the user to provide base images.

Terraform users find Packer valuable for any projects that use VM deployments in a hybrid cloud environment. Properly prepared images will dramatically ease VM provisioning with Terraform. More importantly, cloud-native tooling that may provision instances dynamically, such as autoscale groups or failover routines, will not inform Terraform about their activity. Having a proper VM autoscale group deployed with Terraform still requires a standard image for the cloud to scale.

Alternatives to Packer

Image management has been a challenge for years. Packer is certainly not the first tool to address these difficulties. Tools such as Solaris JumpStart or Red Hat Kickstart have been used to codify VM installation. These can be used in conjunction with Packer to build uniform images across platforms. Packer may use a kickstart to deploy a Linux platform from standard media but then use provisioners to deploy tooling identically across Linux and Windows environments. Docker Compose and Buildah are also modern tools for building specialized container images. Often, specialized community tools such as this can supplement Packer while letting Packer provide a more general-purpose building tool to bring complex mixed environments into one single template. Red Hat Enterprise Linux users have the option of leveraging Red Hat Satellite for platform standardization using a combination of Kickstart, Cobbler, and Puppet.

Historically, simple scripting has been used for early infrastructure as code strategies. If configuration can be scripted, it can be version controlled and used to build and test images captured with either virtualization tools or image tooling such as Norton Ghost.

Installing Packer

Packer is freely available via many options depending on your computer. You can download the full source code at any time via GitHub. In most cases, there is no need to compile your own release. Official binaries are available on HashiCorp's release page: `https://releases.hashicorp.com`. The best way to install is via OS releases:

1. If using Brew for Mac, run the following commands to enable the HashiCorp tap and install Packer:

    ```
    brew tap hashicorp/tap
    brew install hashicorp/tap/packer
    ```

2. For RPM-based Linux distributions, use YUM or DNF to enable the HashiCorp repo for RPM-based Linux using these commands:

    ```
    sudo dnf config-manager --add-repo https://rpm.releases.hashicorp.com/fedora/hashicorp.repo
    sudo dnf -y install packer
    ```

3. For DEB-based Linux distributions, enable HashiCorp's APT repo and install Packer using these commands:

    ```
    curl -fsSL https://apt.releases.hashicorp.com/gpg | sudo apt-key add -
    sudo apt-add-repository "deb [arch=amd64] https://apt.releases.hashicorp.com $(lsb_release -cs) main"
    sudo apt-get update && sudo apt-get install packer
    ```

4. Windows users can use Chocolatey to install Packer using the following command:

    ```
    choco install packer
    ```

The OS packaging contains secure signed binaries that get verified by packaging. Downloading the binaries from HashiCorp's releases page manually requires that downloads be verified manually with checksums before use. This verification ensures the build comes from HashiCorp and doesn't have any compromised code. Never use a Packer release in production without verifying its signature.

HCL versus JSON

It's good to have some basic background on the three coding formats supported by Packer templates. JSON is a descriptive language that uses blocks to declare a data structure. A JSON document may use an optional schema that is a secondary JSON document that lists the structure for the writer to follow. Since version 1.7, Packer actually supports two versions of JSON, so it's important to know how to identify them by file extension when coming across older templates. Legacy templates end in just `.json`, whereas new templates end in `.pkr.json`, and both options use different schemas or styles. HCL is HashiCorp's own syntax, which has a few more features than JSON but also a few limitations:

	JSON	HCL (version 2)
Pros	Widely used across the industry Supports schemas IDE support	Comment support Complex constructs, `for` loops, and here documents Helpful parameters IDE support
Cons	No comments Strict format Lack of constructs: `for` loops and here files, also known as heredocs	No schema support

Table 1.1 – Comparison of JSON and HCL code for Packer

The good news is, Packer supports both HCL and JSON and also has a helpful tool to convert an existing JSON template into an HCL template automatically. HCL may support schemas in the future, but currently, its open format features also help make it more flexible and easier to read than JSON in some cases. Let's start with some examples of both JSON and HCL2 Packer templates. Note there are actually two versions of the JSON schema supported by Packer. The one you use must be reflected in the file extension when you save your template. Legacy JSON templates just end in `*.json` and are supported for existing templates in the community. Newer JSON templates should be written in HCL2.

Example legacy JSON

The following sample is an excerpt from a legacy JSON template used to build an image on VMware. You may encounter these in older examples and Packer still supports them for backward compatibility. Note that some JSON strings contain Go-style templating, indicated by double braces, `{{ }}`. Adding comments is not an option in JSON, so it is difficult to document your code. This code starts with a CentOS 7.8 image, boots it on VMware as specified by a builder, and then uses a provisioner to upload a script and another provisioner to run that script.

JSON schemas provide a way to describe the possible options for a desired JSON document, and can help guide a coder with suggestions, auto-completion, and type checking while building a template. Schemas can also generate WYSIWYG editors, which allow automatic menus and designers for those who don't want to write code manually. Partial community schemas for Packer templates have been written by the author and are available at `https://github.com/jboero/hashicorp-schemas/blob/master/JSON/packer/1.5/template.json`. These schemas are community-driven, not created by HashiCorp engineers. Note that these template samples won't build for you unless you specify a compatible base image. We will actually cover a practical example in the next chapter. A sample template in HCL2 is given in the following code block. We will break down this template line by line in the coming chapters. Optional variables can be declared to help make templates reusable. These definitions look like this and let you define whatever variables you like. Here, there are three variables with default values declared that will be used in builder declarations:

```
variable "base_url" {
  type        = string
  default     = "https://my-source/image.iso"
  description = "URL for our base image"
  sensitive   = false
}
```

Variables in Packer's HCL2 format also offer optional `validation` blocks. This is helpful for limiting what you can assign to the variable. For example, the `base_url` variable in the preceding example is a URL and we want to restrict it to take only values starting with `https`, we can specify this using this `validation` block:

```
validation {
  condition     = substr(var.base_url, 0, 5) == "https"
  error_message = "URLs must start with https"
}
```

There are many variables that come built into Packer for each build or source. These give access to dynamic values, such as the unique identifier for the build, name, and ID of the build resource. This is helpful when you want to inject aspects about the build itself into actions or provisioners performed in each environment. For example, if you want to save the Packer build UUID into the image via a file such as `/etc/packerbuild`, you can reference the `build.PackerRunUUID` variable. A list of the build and source variables can be found in Packer's contextual variable documentation: `https://developer.hashicorp.com/packer/docs/templates/hcl_templates/contextual-variables`.

Builders are plugins used to declare an environment for image building, such as VMware, VirtualBox, QEMU, and Docker. As of Packer version 1.7, templates declare an instance of a builder as a source. In this sample, we declare one builder of the VMWare ISO type with minimal settings to connect our VM. Notice the previous variables are inserted into strings using the {{ }} templating syntax. HCL also supports direct variable usage without strings. A builder says nothing about how your image should be customized. It only tells Packer what kind of environment to run provisioners on to customize your image. Take this example:

```
source "vsphere-iso" "example" {
  iso_url = var.base_url
  iso_checksum = var.base_checksum
  ssh_username = "packer"
  ssh_password = "packer"
  shutdown_command = "shutdown -P now"
  boot_command = [
    "<esc><wait>",
    "vmlinuz initrd=initrd.img ",
    "<enter>"]
  boot_key_interval = "1ms"
  boot_wait = "1s"
  cpus = 8
  memory = 8192
  disk_size = 4000
}
```

Provisioners are the magic of Packer. These are customizations, resources, or scripts that should be run on all of the builders to preconfigure everything you expect in the image. Once all of the provisioners are finished, Packer saves the image as configured in the builder. Here, there are two provisioners. The first is a script called install.sh, which we upload into the builder from a local directory, ./http/install.sh. Then, the second provisioner is a shell command to run that script:

```
provisioner "file" {
  destination = "/tmp/install.sh"
  Source = "./http/install.sh"
  direction = "upload"
}
provisioner "shell" {
  inline = ["sudo bash -x /tmp/install.sh"]
}
```

Packer can be used to build or simply validate this JSON document as a valid template. Note that JSON templates require a root document. Everything is nested within a single set of braces, also known as a code block. This differs from HCL, which requires no root document or block.

Example PKR.JSON

When Packer added HCL2 support, it restructured how templates are structured. There is an additional JSON option that mirrors this HCL2 format. Builders are instead defined as *sources* and then a build job lists which sources you would like to include in the build. It may be a little confusing if you are used to legacy JSON support. Packer will select whether your JSON file uses the legacy or new schema by its file extension. For example, `template.json` uses the legacy schema, as used in the preceding example, whereas `template.pkr.json` would tell Packer to use the new schema of sources. HCL2 is still the recommended way to build new templates, though JSON support still offers some nice automation options for IDEs and UI wizards, which we'll discuss in *Chapter 2, Creating Your First Template*. The equivalent example in `pkr.json` format is listed in the book's GitHub repo: `https://github.com/PacktPublishing/HashiCorp-Packer-in-Production/blob/main/Chapter01/Sample.pkr.json`.

Example HCL

Here, I have taken the previous legacy JSON template and migrated it to HCL2 via Packer's built-in `packer hcl2_upgrade [template.json]` command. I have also added some comments to explain what's happening. HCL supports three comment types: `//`, `/*`, and `#`. I've included examples of all of these types in the following snippet, but it's best to choose one standard and be consistent. HCL has no root object requirement but the structure varies a bit from the JSON version. HCL also supports here docs, also known as here documents, which can help you embed files such as our provisioner script directly into the template. These are often indicated by an `<<EOF` flag or a similar delimiter. The fully converted template with additional comments added manually is shown here. HCL2 can look quite a bit different than JSON. Variables are declared one at a time like in the following example:

```
variable "checksum" {
  type    = string
  default = "087a5743dc6fd6...60d75440eb7be14"
}
```

In addition, each builder is declared separately as a source. Then, a build job lists the sources and provisioners desired:

```
build {
  sources = ["source.vmware-iso.autogenerated_1"]

  provisioner "file" {
    destination = "/tmp/install.sh"
    direction   = "upload"
    source      = "./http/install.sh"
  }
  provisioner "shell" {
    inline = ["sudo bash -x /tmp/install.sh"]
  }
}
```

This HCL2 template provides the same details as the JSON version earlier. It has been automatically converted by Packer and commented to provide more detail. In the next chapter, we will break down every line of this template to explain what each value means in detail.

Summary

This chapter gave a very high-level overview of the HashiCorp tool called Packer. It's actually a very simple tool that delivers powerful results when working with image management at any scale. As a tool, Packer is used when needed rather than as a service that listens for tasks. You can build templates that deliver flexible, scalable, standardized images around the world or in your own private data center. We've explored a few basic sample templates, but the best way to learn how Packer works is to jump in with building your first template, which we'll cover in the next chapter.

2
Creating Your First Template

What is learning a new tool without a *Hello World* app? Let's make a basic Hello World template to break down the basic components of a template. HCL2 is the recommended format to write templates as of version 1.5, but for those who prefer JSON or encounter legacy JSON templates, we will build both HCL and JSON versions. We'll keep this as simple as possible while demonstrating key features in a template that can be built on your own laptop in a coffee shop using free virtualization from Oracle VirtualBox. Coffee will come in handy while you wait for builds to run. We will cover the following topics:

- Hello World template for a local VM

- Breakdown of template components

- Using an IDE to help you write templates

- Applying the VirtualBox builder

Technical requirements

For our Hello World base image, let's choose a simple Linux base image that we will customize with a **Message of the Day (MOTD)** that says **Hello World from Packer**. Then, we will add a few bells and whistles in stages. We'll focus on HCL2 since JSON is considered legacy by the Packer developers. You will need to have Packer version 1.5 or newer installed to try this, and you'll need to either install VirtualBox or select a different hypervisor or builder to use in its place. This build assumes you have an Intel x86_64-compatible PC with VirtualBox installed and sufficient CPU, and at least 8 GB of RAM to run the VM required for the build. Other platforms and architectures such as ARM should work, as long as the template image is adjusted. It doesn't matter whether your operating system is macOS, Windows, or Linux, as long as VirtualBox and Packer are available for your platform.

Hello World template for a local VM

A basic template requires just one builder. We could add data sources to look up external records or use variables to make the template reusable, but we will start basic and refactor it, adding more features later. Each type of builder has different options and parameters, so make sure to check the documented options on Packer's website or via the Packer help menu. Also, we will do a deep dive into builders in *Chapter 6, Working with Builders*.

The Packer documentation is available at `https://www.packer.io/docs`.

Some options are mandatory, and some are optional. We will use comments as appropriate to indicate which is which for the `virtualbox-iso` builder. For convention, I like to use `#` as a comment when disabling a line of code, and `//` or `/*` when writing actual comments. I will break down the meaning of each part for a build of a CentOS Stream base image.

Full copies of these templates can be found at the GitHub repository here: `https://github.com/PacktPublishing/HashiCorp-Packer-in-Production/tree/main/Chapter02`.

Note that there are actually two JSON versions of this template in the repository. Any template file that ends in a `.pkr.json` extension is interpreted by Packer as JSON_HCL formatted JSON. A template that ends in just `.json` without `.pkr.json` is treated as legacy JSON. This may be confusing for new Packer users who find examples on the internet that don't seem to work. Packer 1.5+ uses file extensions to guide its schema parsing. If you save a legacy JSON file as `.pkr.json`, Packer will error about the schema not working because it expects to have a format similar to HCL with sources. Have a good look at all three versions in the repo to understand how they all say the same thing to Packer in different ways.

Breakdown of components, variables, and artifacts

HCL templates rely on the declaration of sources and parameters, which are then invoked in a build declaration. The build block declares a list of sources to use and which provisioners to run on them. In this case, we're defining just one source using the `virtualbox-iso` builder:

```
source "virtualbox-iso" "hello-base" {
```

Note that there are actually multiple builders that support VirtualBox: ISO, OVF, or VM. ISO is the standard way you might install a VirtualBox machine from a DVD image or ISO file. This slightly complicates our installation because we need two steps:

1. Format and install a virtual hard disk.
2. Boot from the new install and finish provisioners over SSH.

Once this image is built into an OVF output, it can be used as a quicker base image to try other build options, since no ISO installation will be necessary. Instead, the VirtualBox OVF builder can be used to rapidly create specialized images. Take a look at the options used for our base image. The order of these parameters in a template does not matter:

```
boot_command = ["<esc><wait>",
   "vmlinuz initrd=initrd.img ",
   "inst.ks=https://github.com/jboero/hashistack/raw/
   master/http/ks-centosStreams.cfg",
   "<enter>"]
boot_wait = "3s"
communicator = "ssh"
```

These options configure how Packer should control the boot process for a new VM. Packer will inject the keystrokes given in `boot_command`. In this case, `boot_wait` is set to 3s before keystrokes are entered. This means Packer will create and boot a new VM, wait three seconds, and then type `boot_command` into the VM's terminal. The boot command is required when starting with a stock image or ISO that does not automatically install to disk on boot. Leaving this out will result in a blank unformatted output image. Special keywords such as `<esc>` are translated to special keys. The escape key is used to bypass the standard ISO boot screen and enter a manual boot command. `<wait>` is used to pause for one second while that screen transition occurs. Finally, the command boots to install a classic Kickstart profile from GitHub.

A Kickstart is used by Red Hat-based distributions, such as CentOS 9 Streams, to declare disk partitioning, system configuration, and pre-installed packages. If the VM does not have access to a publicly hosted Kickstart like the example on GitHub, an HTTP or floppy directory can be configured locally for Packer to inject your content at first boot. If using other distributions or releases that are already distributed as virtual disks, this step becomes unnecessary.

The next attributes set information about the VM to be created for our image. Guest additions aren't necessary for all VirtualBox VMs, but it's a good idea to install them early for integrations anybody might want to use in the future. Here, the CPU count, disk size, and RAM are specified. Note that a disk image won't actually store the CPU or RAM of a machine, and the resources allocated are only used during the build process. The disk will be thin provisioned to save space, meaning the output will only take up as much size as used, rather than the full 100,000 MB:

```
cpus                    = 8
disk_size               = 100000
memory                  = 2048
guest_additions_mode    = "attach"
guest_additions_sha256  =
"62a0c6715bee164817a6f58858dec1d60f01fd0ae00a377a75bbf885ddbd0a61"
guest_additions_url     = https://download.virtualbox.org/
virtualbox/6.1.10/VBoxGuestAdditions_6.1.10.iso
guest_os_type           = "RedHat_64"
```

Note the guest_additions_sha256 option. This is a 256-bit SHA checksum value to verify the integrity of the guest additions ISO. This is for your protection, to make sure that a compromised image isn't swapped into your build after you write the template. All artifacts and downloads should be protected by signatures or checksums. While developing templates and performing several builds to test your results, it's okay to set iso_checksum to none to skip verification, which can take a long time before each build actually starts. You may use a remote URL for your checksum but never do this for production. The remote source can change your checksum and compromise your image at any time. It's best to use static checksums and specific releases rather than the latest image and a remote checksum path. To use a local filesystem path for your ISOs, you can use the file:/// path/to/iso syntax. Packer will not need to download or cache the local image for you. Most likely, repeated builds will be performed either for different environments or for fine-tuning your code as you develop. If this is the case, it will make more sense to use a full ISO or installation image rather than a minimal network boot image, which will result in repeated package downloads over the network. The full media can sit in faster local storage. The cache and network are always a trade-off. If you use a caching proxy, the opposite may be true:

```
#headless            = true
iso_checksum         =
"sha256:fe28038139b6b6e1b85bb690710339eea312cad63281c23d5
6f6da66c5f04a4f "
iso_url              =
"https://lon.mirror.rackspace.com/centos-stream/9-stream/
BaseOS/x86_64/iso/CentOS-Stream-9-20220705.0-x86_64-dvd1.iso "
output_filename      = "hello-world.ovf"
pause_before_connecting = "10s"
```

Comments can be used to disable lines in a template for testing. In this case, the headless flag is disabled by default since the comment blocks it from being set to true. Headless mode runs in the background without showing the user a VirtualBox screen. Some builders have variations of this option. Some builders, such as qemu and libvirt, enable the user to redirect terminal output from the VM directly into Packer output, which is really helpful for logs and requires no GUI option to be available. When building with automation, this will become very important.

Here, we also specify our iso_url parameter or the location to download our boot ISO for VirtualBox. Packer will attempt to download this once and store it inside the local cache in the user's ~/.cache/ packer directory. Be careful to have enough storage available in your home volume for this activity. In the case of the VirtualBox ISO builder, iso_url and iso_checksum are both required, so be sure to obtain or generate a valid checksum beforehand. This option is flexible, accepting md5, sha1, sha256, or sha512 hashes, as indicated in this case by the prefix sha256:. Checksum options may change in future releases. Also, Packer currently has no option for secure signatures, but hopefully will in the future.

Eventually, the output image will need a name. This is hardcoded to hello-world.ovf to keep things simple. If a repeat build is generated and output already exists, Packer will alert you that the output won't be overwritten unless you provide the `-force` option when building. Later, we'll explore variables and template values that allow adding on version numbers or timestamps to keep unique builds without overwriting.

One more important step is the `pause_before_connecting` value of `10s`. If Packer tries to connect before VirtualBox is ready, that may cause trouble. Give VirtualBox some time to prepare and launch our VM before attempting to connect to it:

```
shutdown_command          = "shutdown -P now"
ssh_username              = "root"
ssh_password              = "packer"
ssh_timeout               = "20m"
vm_name                   = "hello-world"
```

Since Packer will perform our provisioners with the SSH communicator, we also have the option of giving a custom command to safely shut down the VM when the build completes. The default option is to halt, which is basically pulling the plug on a live VM without ensuring all disk transactions have been written. Here, we use the safe standard `shutdown -P now` command, and Packer will be logged into the VM as `root` with permissions to safely shut down the VM. There are other options for shutting down a VM. Sometimes, a hibernate can be used to support rapid booting from a VM stored in power state S4. This means booting a VM with the image will just read what was in memory when the machine was hibernated and start up quicker than needing to boot one component at a time.

SSH configuration is important to connect for provisioning. Note that the initial stage of kickstarting an install to the disk must be complete before SSH can connect, and the Kickstart file contains our initial root password, which we will change as soon as possible during final provisioning. Since we are connecting as `root` with SSH for the password, we also will need to configure SSH during kickstarting to allow root logins with a password. This is very insecure for production, and we will cover advanced use cases for secure keys or credentials from a Vault instance later. Do not launch a production VM with root password authentication enabled as it will likely be picked up by brute force attackers or bots.

The name we give this VM is only used during the build within VirtualBox. If VirtualBox already has a VM using this name, then our builder will fail with an error because two VirtualBox VMs can't have the same name. Packer will attempt to remove the VM once a build finishes or errors:

```
vboxmanage = [
      [ "modifyvm", "{{.Name}}", "--paravirtprovider=kvm" ],
]
}
```

This is an optional flag for VirtualBox to set additional parameters on the VM. If there is a custom option in VirtualBox not natively supported by Packer, the builder offers a list of commands to run against the VM before starting it. In this case, there is just one command that translates to `VBoxManage modifyvm {{VM}} -paravirtprovider=kvm`, which will tell VirtualBox to use KVM acceleration within the kernel. KVM is only available on a Linux kernel with KVM support but it will greatly speed up our x86_64 VMs during the build, letting them run nearly at full speed as if they were the host. KVM does not support all guest architectures, so if we add an ARM build later, we may need to adjust its `paravirtprovider` parameter.

This concludes our simple VirtualBox source. If we wanted to build an identical image using CentOS 8 Streams, Ubuntu, or Windows, we could simply add one more source specifying a different image and environment. Furthermore, later, we can also add another source if we want to build an equivalent image in another environment or public cloud. Packer will even build all of them at once, saving a lot of time rather than waiting for one build at a time. We will cover this in more detail in *Chapter 6, Working with Builders*.

For now, the template needs a build declaration. You might think of this as a `main` function for Packer or an entry point for Packer to start its job. The parameters of a build are a list of sources that may be declared inline or referred from outside the build block. In this case, we only have the single source we defined previously. Refer to the sources as `source.{builder}.{name}`, which translates to `source.virtualbox-iso.hello-base`. For each source in this list, Packer will start up the environment it describes and begin to attempt all provisioners affiliated with the build.

There is only one provisioner in this template. Rather than running a script or an Ansible playbook or complex statement, we can start by simply setting the message of the day. Every user who logs into a machine running this image will see our MOTD: **Hello World from Packer**. There are many types of provisioners available, which we'll go on to cover in detail next:

```
// Define a build with our single source builder and two provisioners.
build {
  sources = ["source.virtualbox-iso.hello-base"]
  provisioner "file" {
    destination = "/etc/motd"
    direction   = "upload"
    content     = "Hello World from Packer."
  }
}
```

Now, if your machine has access to the internet to download all ISO resources in the template as well as the Kickstart file, you can install VirtualBox and let Packer perform its build via the simple `packer build` command. Packer naming conventions were changed when HCL2 support was added. Template names should take the format `{NAME}.pkr.hcl` for HCL2 templates or `{NAME}.pkr.json` for JSON templates. Legacy JSON templates are still supported and Packer will recognize them by it not having the double extension `pkr.json`, so pay close attention to the file extension used:

```
$ packer build hello.pkr.hcl

virtualbox-iso.hello-base: output will be in this color.
==> virtualbox-iso.hello-base: Retrieving Guest additions
==> virtualbox-iso.hello-base: Trying https://download.virtualbox.org/
```

```
virtualbox/6.1.10/VBoxGuestAdditions_6.1.10.iso
==> virtualbox-iso.hello-base: Trying https://download.virtualbox.org/
virtualbox/6.1.10/VBoxGuestAdditions_6.1.10.iso?checksum=62a0
c6715bee164817a6f58858dec1d60f01fd0ae00a377a75bbf885ddbd0a61
==> virtualbox-iso.hello-base: https://download.virtualbox.org/
virtualbox/6.1.10/VBoxGuestAdditions_6.1.10.iso?checksum=62a0c6715be
e164817a6f58858dec1d60f01fd0ae00a377a75bbf885ddbd0a61 => /home/
jboero/.cache/packer/486bd274e88f530971538017478eeb91b031c75b.iso
==> virtualbox-iso.hello-base: Retrieving ISO
==> virtualbox-iso.hello-base: Trying http://lon.mirror.rackspace.com/
centos-stream/9-stream/BaseOS/x86_64/iso/CentOS-Stream-9-late
st-x86_64-dvd1.iso
==> virtualbox-iso.hello-base: Trying http://lon.mirror.rackspace.com/
centos-stream/9-stream/BaseOS/x86_64/iso/CentOS-Stream-9-late
st-x86_64-dvd1.iso?checksum=sha256%3A5af0d4535a13e8b1c5129df85f6e8c
853f0a82f77d8614d4f91187bc7aa94d52
. . . .
```

You may also run `packer build` on a directory. In this case, Packer will combine all .hcl files in the current directory, which helps break out code into multiple files or use symlinks as includes. Do not mix JSON and HCL templates in the same directory or Packer will get confused. Strategy ideas for how to take advantage of this will be described in future chapters as requirements get more complex.

Note how helpful Packer is to colorize the output. With one builder, this may not be very helpful, but when using multiple builders in parallel, this becomes very useful to distinguish outputs. If using automation, it's important to save the colorized output or it will be difficult to distinguish builders.

Using an IDE to guide templates

What would a JSON version of our first template look like? JSON is important because a lot of source material from the community and GitHub is still written in legacy JSON. In fact, JSON has some advantages and disadvantages, so you may want to start with JSON when writing your template and convert it to HCL later. There is no tool within Packer to convert HCL2 to pkr.json or legacy JSON, so when a JSON option is required, it's best to start with pkr.json syntax and not legacy JSON. Only legacy JSON to HCL2 templates can be migrated with Packer's built-in tool. This can be installed from GitHub or via the go command:

```
go install github.com/hashicorp/hcl/v2/cmd/hcldec@latest
```

Schemas are very helpful when building JSON templates. A JSON schema is simply a specialized JSON document that declares the desired format of another JSON document. Schemas are not available from the Packer engineering team, but Community versions are available on GitHub: https://github.com/jboero/hashicorp-schemas/tree/master/JSON/packer.

Schemas can be used within an IDE or GUI framework to provide guided designers for a Packer template. Autocompletion will offer defaults, type-safe suggestions, and help highlight any errors before Packer finds them. Simply adding a root value specifying the schema to use inside Visual Studio Code is enough to let the IDE guide a developer to the perfect template.

Here is an example: `"$schema": "http://example/schema.json"`.

```
{} manifest.pkr.json 1 ●
home > jboero > {} manifest.pkr.json > [ ] builders > {} 0
   1    {
   2          "builders": [
   3                {
   4                      "type": "virtualbox-iso",
   5                      "headless": true,
   6                      "disk_size": 40000,
   7                      |
   8                }          ⚙ acpi_shutdown
   9          ]            ⚙ boot_command
  10    }                ⚙ boot_keygroup_interval
  11                       ⚙ boot_wait
                           ⚙ bundle_iso
                           ⚙ communicator
                           ⚙ cpus
                           ⚙ disable_shutdown
                           ⚙ export_opts
                           ⚙ floppy_dirs
                           ⚙ floppy_files
                           ⚙ floppy_label
```

Figure 2.1 – Schema autocompletion in Visual Studio Code helps guide template design

UI frameworks such as JSON Designer and other tools can be used to build a simple template designer where code is not preferred. Instead, a simple menu form will be presented to the user as a view to the JSON underneath. As JSON templates require a root document, simple tools such as JQ or standard JSON APIs can also be used to build or inspect templates. Often, teams that have existing tooling built on JSON standards continue to use JSON with Packer and do not plan to migrate to HCL. The fact that JSON schemas tell the user what fields are required and which are optional is reason alone to give them a try as you learn Packer.

Applying the VirtualBox builder

VirtualBox might not be considered a production hypervisor, but it is actually very helpful when learning Packer on your own. VirtualBox works on most common operating systems and gives you the option to connect and troubleshoot a VM during an image build. Note that the builders that support VirtualBox offer a *headless* option, meaning you will not launch VMs in a GUI during a build. This is handy if you are working remotely via SSH or using automation.

A caveat to the VirtualBox builder is lack of access to TTY from your Packer session. This means the machine's text output won't directly show up in Packer output. Provisioners executing over SSH have the handy `ssh_pty` option, which allows you to capture SSH output. For the rest, you should probably use `headless = "false"`, which is the default and lets you view the VM in VirtualBox. Later, we will cover other builders such as QEMU, which can actually output all of a VM's TTY console directly to Packer.

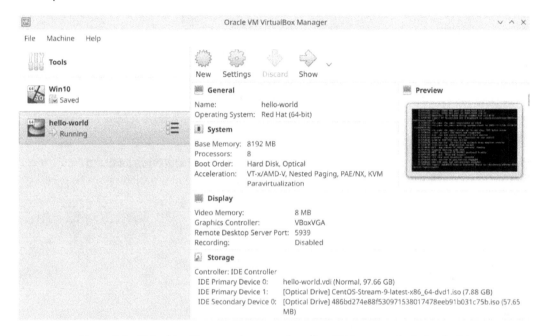

Figure 2.2 – VirtualBox Manager shows our headless VM temporarily during build

If a VirtualBox build is successful, Packer will place the resulting image into your current working directory at `./output-hello-base/`, which uses the output name declared in each builder source. The output is an **Open Virtualization Format** (**OVF**) bundle. This format is an industry-standard format compatible with many hypervisors, including VirtualBox, VMware, Red Hat Virtualization, Amazon, OpenStack, and more. The output contains at least two files. The first is an XML file with generic machine descriptions including CPU, RAM allocations, and network config. The second is the VMDK disk output from the VM.

VMDK is another common OVF alternative format used to capture disk images. An image may be thin-provisioned or thick-provisioned, such that only the used storage is retained. Usually, for packaging images, thin provisioning is used to not waste extra space. VMDK is a pretty common disk format, but it does not support native compression much like other image formats, as we'll see later. Packaging these files together into a `tar` archive or `gzip`-compressed `tar` archive results in an **Open Virtual Appliance** (**OVA**), which can be imported into many hypervisors as a template for provisioning VMs. Be sure to add the `.ova` extension if you do this manually to distinguish it from a standard `tar` archive. If you find OVA appliances around the internet or in repositories, you may simply extract a copy of the archive to inspect or edit them. Now that we've covered the basics of a single builder, we will expand into other builders and key options that will help you along the way in *Chapter 6, Working with Builders*.

Summary

In this chapter, we showed a basic template to provide us with a first base image for Packer to build further images. We selected CentOS 9 Streams as a Linux distribution to simulate enterprise environments, but next, we will add other options to our template, including other Linux distributions and Windows. Templates can be written in either HCL2 or JSON. HCL2 is the recommended template format for new projects, but JSON templates are widely available on the internet and easily converted to HCL2 if necessary. JSON can still help when automation is used to build templates, but from here forward, we will focus on writing example templates in HCL2.

Simple templates like the one in this chapter cut a lot of corners and hardcode a lot of attributes, but in the next chapter, it will be apparent how limiting that can be and how variables can be used to simplify templates and improve reusable attributes.

3

Configuring Builders and Sources

In the previous chapter, we dove straight into Packer with a detailed but simple example of using the VirtualBox ISO builder, which should be freely available on most environments for you to try out your first image build. We generated a simple OVF without much complexity.

In this chapter, we will expand on builders and explain the most commonly used local and remote options, including their configuration options. First, we'll describe the full set of VirtualBox builders for the example we covered in the previous chapter. Then, we will explore more common builder options, such as VMware, and public clouds, such as AWS, Azure, and GCP. Note that if you add another builder declaration to your template, it will be built in parallel by default. If we add a VMware builder to our template, Packer will try to build VMware and VirtualBox images at the same time. You can also declare multiple builders of the same type, such as VirtualBox ISO builders for RHEL, Ubuntu, and Windows. These builds will be attempted in parallel unless the parallel options are adjusted when you build.

The aim is to establish a consistent set of standards, also known as *gold* or *golden* system images, with common attributes that are needed for your organization. These specialized images can be built with specific applications and use cases, such as databases, web servers, or autoscale groups. These gold images should cover the necessary combinations of operating systems and image formats. We will aim to create both OVF and QCOW2 images from Windows Server 2022, CentOS Streams, and Ubuntu Server operating systems. We will cover a variety of builder options to accomplish this:

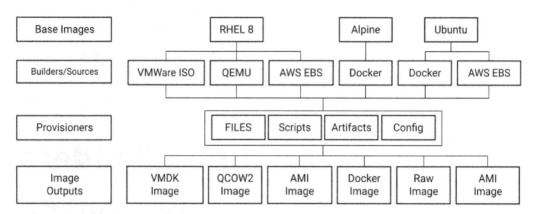

Figure 3.1 – Example structure with common base images sharing common provisioners

Along the way, we will introduce some important tools to help you manage multiple builds. The biggest helper here is the option to use variables and locals. When multiple builders all need to specify common parameters such as ISO URLs or SSH usernames and passwords, it's really helpful to define repeated values as variables so that they only need to be set in one location or can even be used as parameters when calling build jobs. We will cover this and more in the following topics:

- Simplifying your template with variables
- Utilizing local system builders
- Adding cloud builders

Technical requirements

In this chapter, we will cover cloud providers that require a paid account to follow the example code. If you don't have access to a cloud account, we will also cover local builders, including QEMU, which can be used for free but requires setting up a local QEMU environment on your PC. Check the QEMU documentation or your Linux distribution guide for instructions on setting this up: https://www. qemu.org/docs/master/.

Otherwise, it may be possible to contact the cloud providers for a trial to follow along with the cloud source code examples.

Simplifying your template with variables

Variables can be declared in HCL2 in a very similar way to how Terraform declares variables: create a variable block, give it a name, give it a description and type (optional), and specify an optional default value. Here, we will create a variable called iso that defaults to a download URL. This may be referenced in the code directly as `var.iso` or it can be used with standard shell interpolation in strings, as in `${var.creds.user}`, which helps with templating:

```
variable "iso" {
  description = "The path to your ISO."
  type = string
  default = "https://fqdn/media/base.iso"
}
```

Local variables differ from normal variables in that they can't be assigned as parameters. Locals offer a simple way to declare reusable values in your code without allowing the runner to specify them as parameters. You can think of variables as public parameters and locals as private variables, which cannot be set without you changing the code:

```
locals {
  org = "myorg"
  zone = "{var.region}-a"
}
```

Variables can also have an `object` type for their structure. We will likely need a variable for provisioner credentials in the form of a username and password. Rather than two variables, we can simplify this with an `object` variable. In this case, `creds` is a variable of the `object` type with two strings. It defaults to `{ "user" = "user", "pass" = "pass" }`. We can specify these during the build so that we don't reveal credentials in our code. Here is an example:

```
variable "creds" {
  description = "User and password credentials for provisioners."
  type     = object({
    user = string
    pass = string
  })
  default = {
    user = "user"
    pass = "pass"
  }
}
```

To save ourselves from re-keying a username and password for every builder, let's use a variable. Later, we will demonstrate how to look up variables externally or even import secret variables we don't want to reveal in logs. For now, let's go ahead and try adding some local builders to our current image.

Utilizing local system builders

Packer isn't always about cloud or remote system images. Sometimes, it's helpful to build local images for testing or local hypervisors. The build performance in local environments is often better and may save you time when developing. Here, we will cover a few common options for local builds before moving on to cloud images.

Using VirtualBox builders – ISO, OVF, and VM

In addition to the ISO builder, which boots a VirtualBox VM from a read-only ISO image, you may use an existing OVF, or even an existing VM that exists in VirtualBox. The OVF builder is particularly useful as it lets images quickly be built from the OVF output of another Packer build. Let's explore the options for each of these. Note that if you're using JSON with the JSON schemas enabled in your editor, these options should be annotated and guided with optional default values. One thing we didn't cover is configuring the VirtualBox plugin explicitly in your Packer configuration block. Each plugin can be configured in a special block within your template. This allows you to specify a custom source or require certain versions of a plugin:

```
packer {
  required_version = ">= 1.7.9"
  required_plugins {
    virtualbox = {
      version = ">= 0.0.1"
      source = "github.com/hashicorp/virtualbox"
    }
  }
}
```

In this case, we're explicitly telling Packer where to get the source code for our plugin. We can use this to override the version that's built into the current Packer binary or force a specific version to be used. Some plugins also supply their own environment or variables, such as cloud credentials or API tokens. Any time you set or change a plugin configuration like this, you should run `packer init` so that Packer ensures the correct plugins are downloaded and ready to use. The VirtualBox plugin covers all three builders involving VirtualBox integration: ISO, OVF, and VM. Here, we cover a few examples while assuming that the same provisioners we used with the ISO builder are used in the OVF and VM samples as well.

We've already covered ISO and how to build from standard optical media images, so let's learn how to use the OVF and VM variants. If you're using VirtualBox to build on top of an existing or base image in OVF format, all that's necessary is to change the source type and download from `iso_url` to `source_path`, which will let the base image be extended into a new custom image. Typical use cases would be to install an application or web service on top of the base image. The OVF builder will act much faster than an ISO builder because there is no need to install the base OS and format storage. So, it's good to start with a common OVF base if one isn't offered by your operating system's distribution:

```
source "virtualbox-ovf" "ovf-example" {
  source_path = "base-source.ovf"
  // More parameters
}
```

The VM option is used when there is an existing VM inside VirtualBox that should be used to build your output. It uses snapshots to save a restore point for changes rather than disposing of the VM after the build. It can be useful for updating base images without building from scratch since the VM won't be destroyed between builds. Building can also result in unexpected behavior if your provisioning is not idempotent. This means that your provisioners affect the VM state in a way that they can't be reliably run again without breaking. Often, the most consistent way to reproduce any environment is from scratch, so use the VM builder option only when necessary. Disposable infrastructure expects a clean slate and the ability to recreate an environment from scratch when necessary. Here is a snippet that uses a selected VM and snapshot to generate a new snapshot and output an image from it:

```
source "virtualbox-vm" "vm-example" {
  vm_name = "your-existing-vm"
  attach_snapshot = "snap_01"
  target_snapshot = "snap_02"
  force_delete_snapshot = true
  keep_registered = false
  skip_export = false
  // More parameters
}
```

Here, we attach to the existing VM, called `your-existing-vm`, and restore the `base_01` snapshot before running whatever provisioners are included in the build. When finished, Packer will take another snapshot of the `base_02` VM and save an OVF image from it. After a successful export, Packer will delete the snapshot since we set `keep_registered` to `false`. Optionally, the snapshot can be saved, though several layers of snapshots can be difficult to manage and may affect performance. If a snapshot called `base_02` already exists, we will overwrite it by setting `force_delete_snapshot` to `true`. This defaults to `false`, and Packer will exit with an error before it overwrites the target snapshot.

You can continue to learn to use Packer solely based on these three VirtualBox builders if they are your preferred option, but you may benefit from some of the advanced features in the VMware and QEMU builders, which we'll cover next.

Using VMware builders – ISO, VMX, and vSphere clone

The OVF or OVA output from VirtualBox builders is also compatible with VMware. However, if you would like to generate your images within VMware, you have a few options. VMware has an ISO builder just like VirtualBox has, but the ISO builder can be used both with local VMware or remote ESX servers. There is also a dedicated vSphere builder for VMware clusters, which will be more common in enterprise production environments. There is also an optional path to the VMware Fusion binary if you use VMware Fusion.

The VMware plugin can be configured via the Packer block. In this case, there are two different plugins. vSphere and VMware are separate plugins with somewhat overlapping features. Both plugins offer ISO builders, so make sure you configure the appropriate plugin. As we are focusing on local builders, let's talk about the VMware ISO and VMX options first. The ISO and VMX builders are almost identical to the VirtualBox ISO and OVF builders, respectively. VMX is VMware's standard for passing around virtual machines, while VMware also supports OVF. There are widely available tools for managing and converting OVF and VMX archives. The most common one is a command-line application called **ovftool**.

Let's start by adding VMware support to our existing template so that VMware can use our existing provisioners to generate a VMware image. These will require a licensed version of VMware Workstation or VMware Fusion, or at least the VMware `vmware` and `vmrun` binaries to be installed in the path. The VMware plugin runs with local `exec` commands to interact with VMware instead of using a direct API. This contrasts with the vSphere plugin, which uses API calls to an external VMware vSphere instance:

```
source "vmware-iso" "base-example" {
  iso_url = var.streams_iso.url
  iso_checksum = var.streams_iso.shasum
  ssh_username = "user"
  ssh_password = "pass"
}
```

Does this look familiar? Most of the VMware builders behave very similarly to VirtualBox builders. In this case, we're using our new variable, `streams_iso`, so we don't need to hardcode our URL or `sha256sum` in each ISO builder. Some builders allow a list of URLs for `iso_urls` instead of a single `iso_url`. In this case, Packer will try the URLs in order until it finds one it can fetch. In this case, it might be useful to change the variable object definition for the URL to a list of strings.

What about vSphere builds? If your team uses vSphere, you have the benefit of building images directly into your shared environment. The fundamentals of VM setup, such as CPU, RAM, and storage, are all fairly simple. The difference is configuring a vSphere connection and credentials. For these builders, a licensed vCenter must be configured and running. Let's say we have a vSphere cluster and a vCenter server at `vcenter.your.domain`. A sample builder configured to build our same base image from ISO looks like this, though valid credentials must be substituted:

```
source "vsphere-iso" "base-example" {
  vcenter_server = "vcenter.your.domain"
  username = "vcenteradmin"
  password = "yourvcenterpass"
  insecure_connection = true
  datacenter = "yourdc"
}
```

Setting `vcenter_server` tells Packer where to direct our request. Instead of using an `exec` provisioner locally, it will use the vSphere API to coordinate with the external cluster. This is also how the cloud builder options work, which will be covered in the next section. The credentials given must have permissions to manage resources and storage pools required for the build. Again, it's not wise to leave sensitive credentials in a template like this, so normally, variables should be used. Usually, a vCenter will be protected with valid TLS. However, if a self-signed or invalid certificate is used, `insecure_connection` can be set to `true` to skip TLS verification. It would not be wise to use this option in production.

The vSphere clone builder is very similar to the VirtualBox VM builder. It takes an existing VM and optionally uses snapshots to provision it before saving an image. In this case, taking a snapshot is not the default, so beware of mixing the `vsphere-clone` and `virtualbox-vm` builders:

```
source "vsphere-clone" "base-example" {
    vcenter_server        = "vcenter.your.domain"
    username              = "vcenteradmin"
    password              = "yourvcenterpass"
    template              = "test"
    vm_name               = "vsphere-base"
    create_snapshot       = true
    snapshot_name         = "packer_snap"
}
```

This is a powerful way to build complex image hierarchies with vSphere privately. Start with the ISO builder to create base images for your Linux and Windows systems. Base images from ISO can be configured to harden the OS with **single sign-on** (**SSO**), or attached to LDAP and Kerberos and Windows AD and external log aggregators to ensure that other images already have the basics installed for your environment. Then, you can build further images on top of those with `vsphere-clone`.

Using the QEMU builder

The QEMU builder is arguably the most powerful builder available to Packer. QEMU allows you to specify the `qemu-system` binary that's used during a build. This means you can cross-build images for other architectures that are supported by QEMU, including ARM, AArch64, IBM Power, S390/X, and even microcontrollers. QEMU is probably the most flexible local builder available, even though it's not the most commonly used. QEMU's KVM acceleration requires supported platforms, so builds are generally performed on Linux:

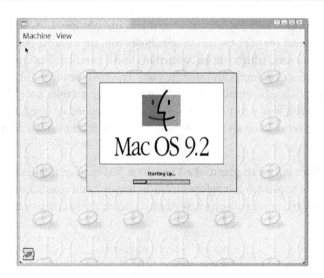

Figure 3.2 – Legacy images created with Packer and qemu-system-ppc

The QEMU builder also has one unique feature, which is TTY forwarding. The console output from the VM's default TTYS0 can be output into Packer logs. You can record a VM building line by line, even in headless mode, which is helpful for troubleshooting. To include TTY output, you must use the PACKER_LOG=1 environment variable and specify -nographic in qemuargs. Beware that the QEMU output may potentially output sensitive values in Packer logs:

```
source "qemu" "base-x86_64" {
  accelerator = "kvm"
  headless = true
  qemuargs =[
    // Set nographic for console to Packer.
    ["-nographic"],
    // Or have QEMU write its output to a file.
    ["-serial", "file:vm.out"]
  ]
  // More parameters
}
```

Here, we have a sample builder that will start a QEMU VM locally and include VM terminal output in Packer logs. This includes all BIOS and boot information, along with timestamps, which is helpful to record. Alternatively, you can have QEMU dump its serial output into a file. This can be helpful but it will not appear in line with Packer's output to help troubleshoot timing issues:

```
2022/02/01 16:32:44 packer-builder-qemu plugin: Qemu stdout: Booting
from DVD/CD...
2022/02/01 16:32:44 packer-builder-qemu plugin: Qemu stdout:
2022/02/01 16:32:44 packer-builder-qemu plugin: Qemu stdout: ISOLINUX
6.04  ETCD Copyright (C) 1994-2015 H. Peter Anvin et al
```

In addition to the usual VM specs such as CPU, RAM, and storage, we can control various aspects of QEMU, including the accelerator that's used. KVM provides a major boost to architectures that support it on your QEMU host. Packer will attempt to use KVM by default or switch to the generic virtualized **Tiny Code Generator** (**TCG**) where KVM is not available. KVM generally supports native virtualization that matches the host's hardware, plus architectures that have been supported by the QEMU developers. For architectures that don't support native KVM acceleration, it's wise to disable acceleration for a fully emulated environment. This won't be nearly as quick as the KVM build but the job should be done without errors or architecture issues. You can use QEMU to emulate a wide variety of platforms, from embedded or IoT devices, Raspberry Pi to S390/X mainframes, and IBM Power PC, but often, TCG must be used to emulate these environments at slow speeds. QEMU even supports old m68k Mac processors if you want to use Packer to play with legacy 68k or PPC Mac images. Let's look at another example that builds a base image for AArch64:

```
source "qemu" "base-aarch64" {
  accelerator = "none"
  qemu_binary = "qemu-system-aarch64"
  machine_type = "virt"
  // More parameters..
}
```

In this example, we override the QEMU builder's `qemu_binary` from its default of `qemu_system_x86-64` to run an AArch64 VM. This can be used to prepare images for ARM environments or AWS Graviton. Note that any source media that are used, such as ISO or cloud images, will need to be the appropriate variants, as the standard x86_64 ISOs for Linux or Windows can't be used on incompatible architectures. Note that if your build host is an AArch64 system, it can use KVM acceleration. If you're running on x86_64, you will need to either find a supported acceleration option or play it safe by using `none` for full virtualization. It won't be fast but it will fully emulate the CPU for you.

The output of a QEMU build is a qcow2 or a RAW image. qcow2 is similar to the VMDK or VDI format used by VirtualBox and VMware and it is supported by most open virtualization options, including KVM, OpenStack, Red Hat Virtualization, Proxmox, and more. It supports compression and thin provisioning, which can be useful for saving space. qcow images can also be mounted as local storage in a Linux environment via the `qemu-img` tool and kernel module. This makes it easy to copy a qcow image to another source or even a raw device. This can even be used to turn a Packer build into a PXE network boot image. It's also possible to mount the images in *read-write* mode if you need to make changes outside of Packer – though it wouldn't be wise to do so without automation of some kind.

What about microcontrollers? Microcontrollers are the common low-power chips that power things such as smart appliances, digital clocks, and IoT devices. A lot of people don't realize that `qemu-system-svt` can emulate everything from Arduino to mass-produced, industrial microcontroller devices. Microcontrollers don't run a full operating system – just your application. You might think of them as the hardware equivalent of a container since you don't need to worry about an OS or other background processes. As such, they have minimal memory (16-48k typically) and storage and require minimal power – usually batteries in the field. Common use cases for microcontrollers include anything with a control panel that uses buttons and lights that may or may not be connected to your home Wi-Fi or even a Helium IoT network. These can include the following:

- Smart washing machines that notify you when your load is done

- Smart mouse traps that notify your phone when a trap has been tripped

- Smart buoys or weather monitoring devices powered by solar

- Smart fridges or devices warning you when the power goes out

- Smart oven devices warning you they've been running too long

Imagine a use case where you have a mobile app connected to your cloud service, all being fed data from microcontrollers in a global network of smart appliances. All three of these can be built and tested within a single Packer template using the QEMU builder. If your microcontroller image requires configuration or updates based on your backend application, all three images can be built together using Packer:

- Using a backend application on a cloud VM

- Using a mobile microservice on a Docker container

- Using a mobile device or Android tablet

- Using a virtual smart appliance microcontroller

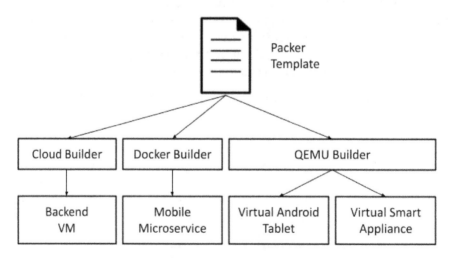

Figure 3.3 – A full stack application from one Packer template

QEMU can even emulate industrial **Programmable Logic Controllers** (**PLCs**). These expensive controllers typically operate automated assembly lines or heavy equipment, from CNC robots to carnival rides.

Not only can QEMU build these complex mixed architecture images but QEMU itself can also be run on just about any architecture. A simple Raspberry Pi device can run any of the builds from this book or its sample repository, which is helpful if you're a student or hobbyist trying to learn how to use Packer.

Now that we've covered the most common local builder options, we will cover common public cloud builders that can help migrate our sample image to the cloud. Cloud builders can be added without changing the way local images are built so that there is a single workflow for all environments.

Adding cloud builders

Building base images locally during development can save a lot of time and money. Waiting for cloud resources to be provisioned can be fairly quick, but usually not as quick as a well-tuned local resource. Also, repeated attempts to provision while addressing minor mistakes during development can generate a significant cloud bill. Once development is complete, you may want to push production images directly through cloud builders.

Providing your cloud credentials

Most cloud builders use similar credential configurations. Static credentials can be defined in your code. As usual, it isn't wise to store these raw credentials in code. Environment variables can be used for credentials stored outside of the template code. These would need to be set in your build environment or Packer automation pipeline. Even better, we will cover how **HashiCorp Vault** can issue short-term credentials to Packer templates so that any leaked logs with credentials only show old, revoked credentials. HashiCorp Vault can act as a dynamic secret generator to give Packer a short-term cloud credential, which is good for a short-term lease. Once the lease expires or is revoked by your build, the credential will be revoked and invalidated so that nobody can use it. HashiCorp Vault itself is outside the scope of this book, but with access to a Vault, the address and token credentials can be fetched directly into a local variable for use throughout your template:

```
export VAULT_ADDR=https://yourvault:8200
export VAULT_TOKEN=[YOURTOKEN]
```

Once the Packer process has these set to a valid Vault and token, the `vault()` function can be used to request secrets from the Vault, including dynamically generated cloud credentials. The `vault` function requires a path and a key. You must fetch the entire secret as a key. Fetching the user or password separately would create two credential pairs with mismatched passwords. It is helpful to declare these as a `sensitive` local variable so that they won't be leaked in logs:

```
local "creds" {
  expression = vault("/aws", "creds")
  sensitive = true
}
```

Now, these can be used in place of the sample `var.creds` object we created earlier. Instead, you can reference your credentials as `local.vaultcreds.secretkey`, for example. Configure Vault with the shortest term lease to finish your cloud builds, but not so short that they time out and are revoked during the build.

Using the AWS builders

AWS support is provided in the Amazon plugin, which provides quite a few builder options.

Here is a list of AWS builder options:

- `amazon-ebs`
- `amazon-instance`
- `amazon-chroot`
- `amazon-ebssurrogate`
- `amazon-ebsvolume`

Amazon EBS is the best place to start since it simply selects an AMI, builds a VM with it, and saves an EBS. A sample of this is included here. Note that the `tags` object is common across most cloud offerings. In this case, we will use a special function inside HCL2 called `formatdate`, which will automatically add a tag to our image specifying the date it was built:

```
source "amazon-ebs" "base-aws" {
  region        = "us-east-1"
  source_ami    = "ami-fce3c696"
  instance_type = "t2.micro"
  ssh_username  = var.creds.user
  ssh_password  = var.creds.pass
  ami_name      = YOUR-AMI-ID
  tags          = {
    owner       = "boero"
    timestamp   = formatdate("YYYYMMDD")
  }
}
```

AMI IDs can be tricky to track. There is a very helpful data source within the Amazon plugin for looking up AMI IDs by name or search. Data sources can help dynamically look up external sources during the build. In this case, you can search for AMIs with wildcard support and select the latest result:

```
data "amazon-ami" "rhel-latest" {
  filters = {
    virtualization-type = "hvm"
    name = "rhel/images/*"
    root-device-type = "ebs"
```

```
    owners = "309956199498"
  }
  most_recent = true
}
```

This code snippet uses the Amazon plugin's `amazon-ami` data source to look up AMI attributes by name. You can also select your AMI ID manually. For testing purposes, the `most_recent` option can be helpful, but keep in mind that builds may fail since underlying AMI updates affect builds. To reference this data source in a template, use a notation like this:

```
source_ami     = data.amazon-ami.rhel-latest.id
```

If you would like to upload your local image directly into your AWS storage, AWS supports importing OVA, OVF, VHD/X, and VMDK formats. This may save compute resources at the cost of ingress network costs for the upload. This should only be done by advanced users or distribution managers, as tuning the build and kernel for AWS hypervisors can be tricky. Most cloud providers provide a tool such as `rsync` to synchronize diffs in local data, which helps minimize data transferred between builds. The AWS option is called `DataSync`.

In this section, we covered the basics of AWS builders. We will use the EBS builder when we explore provisioners and different postprocessing options in the next chapter.

Using the Azure builders

Microsoft Azure is a bit simpler than AWS when it comes to Packer. The Azure plugin only has three builders: **Azure Resource Manager** (**ARM**), chroot, and DTL or DevTest Lab. ARM is the most common builder, similar to Amazon's EBS, whereas DTL is great for building out labs for others to use. The chroot builder is particularly interesting. As the name implies, since this is a chroot environment, this builder does not require a new VM for every build. It can be built in an already running instance, which saves quite a bit of time during builds. Azure credentials are provided directly to the builder, which doesn't support environment variables such as the Amazon plugin:

```
source "azure-arm" "basic-example" {
  client_id = "{YOURCLIENTGUID}"
  client_secret = "{YOURSECRET}"
  resource_group_name = "yourpackerrg"
  storage_account = "yourstorage"
  subscription_id = "{YOURSUBSCRIPTIONGUID}"
  tenant_id = "{YOURTENANTGUID}"

  os_type = "Linux"
  image_publisher = "Canonical"
  image_offer = "UbuntuServer"
  image_sku = "20.04.4-LTS"
  location = "West Europe"
```

```
    vm_size = "Standard_A1"
}
```

This simple example shows the minimum configuration needed for an Azure builder. HashiCorp Vault can be used to generate short-lived credentials for Azure builds using the same technique described for AWS:

```
local "azure-vault" {
  creds = vault("/secret/azure", "packer")
}
```

Azure users will likely be building Windows VMs with Packer, but as this chapter focuses on builders, this example deploys a Linux VM. In the next chapter, we will cover provisioners in more detail and how to switch between the SSH and WinRM communicators for connecting to Windows VMs.

Azure's chroot and DTL builders can be used to help scale out large builds but require more configuration to use. DTL requires a DevTest Lab to be enabled in Azure before use. Chroot requires storage to be pre-created before the builds. This storage can be reused, which helps save provisioning time.

The Azure plugin also provides a DTL provisioner, which is a helpful way to add a DTL resource natively to a VM without using a communicator at all. This would be very helpful at scale if you are only including Azure resources in your template since it only supports Azure builders.

Using the GCP builders

Google Cloud Platform (GCP) support is fairly simple. There is only one builder and authentication happens automatically when running Packer from GCP or GKE. If you're running Packer locally, there are a few different ways to specify credentials, including Vault support, which is similar to the Vault support for AWS and Azure. By default, Packer will use the same configuration as set by the gcloud CLI tool, which is helpful. Just make sure permissions on your credentials file are minimal so that nobody else can access your credentials. For maximum security when using any of these cloud options, use Vault for unique and temporary credentials for every build:

```
source "googlecompute" "basic-example" {
  project_id = "YOURPROJECT"
  source_image = "debian-9-stretch-v20200805"
  zone = "us-central1-a"
  // More parameters
}
```

There are four options to authenticate the googlecompute plugin:

- **Option 1**: Auth by access token:

  ```
  access_token = "{YOURACCESSTOKEN}"
  ```

- **Option 2**: Auth by account file:

  ```
  account_file = "/path/to/account.json"
  ```

- **Option 3**: The GOOGLE_APPLICATION_CREDENTIALS environment variable

- **Option 4**: Put the Google account file in your home directory

Use whichever option you require and be sure to check the plugin documentation page for the latest information: https://developer.hashicorp.com/packer/plugins/builders/googlecompute

Exploring other cloud builders

Other builders are available in Packer via community plugins. In this section, we'll look at some of the known plugins that support builders you may be interested in. Check the Packer documentation for details on these if you need them.

Here are the various public cloud builder options:

- AliCloud
- DigitalOcean
- Hertzner Cloud
- Huawei Cloud
- IBM Cloud
- Naver Cloud
- Oracle Cloud
- Tencent Cloud
- Joyent Cloud (Triton)
- Yandex Cloud
- Linode

Here are the various private cloud and hypervisor builder options:

- HyperV
- OpenStack
- OVirt
- ProxMox

Be sure to check the current plugin support on Packer's community page and GitHub if you are looking for support for other solutions. A lot more exist in the community and more options will be added as Packer matures. If you need to write builders or plugins, we will cover how to do that in *Chapter 13, Developing Custom Plugins*.

Building containers

Docker, LXC, LXD, and Podman are supported via Packer plugins. At first, it may seem counter-intuitive to use Packer for building containers. Docker does a fairly good job of making container images. The nice part about adding a container builder to your template is that it allows you to package your applications and config in containers at the same time as cloud or VM images. In addition to the builder, the Docker plugin has some very helpful postprocessors to manage the container registry lifecycle:

- `docker import`
- `docker push`
- `docker save`
- `docker tag`

You may have Packer build and push containers to your private registry as part of your automation. You may also automate tagging with Packer's variables or useful HCL date/time functions. Here is an example:

```
source "docker" "base-docker" {
  image = "ubuntu"
  commit = true
    changes = [
    "VOLUME /local /container",
    "EXPOSE 80 443",
    "ENTRYPOINT /usr/bin/start"
  ]
}
```

Here, we took the standard `ubuntu:latest` image and changed it a bit. Docker will connect directly to the container to run any provisioners declared. There is no need for the SSH or WinRM communicator or authentication credentials.

Postprocessors are applied by a source after the build. For example, if you want Packer to tag and push the image to your registry, add this postprocessor block to your Docker builder source:

```
post-processors {
  post-processor "docker-tag" {
    repository =  "yourrepo/yourimage"
    tags = ["1.0-packer"]
```

```
  }
  post-processor "docker-push" {}
}
```

The process of adding a postprocessor to a source using the Docker builder results in an automatic tag and push to the image registry provided. The process running Packer needs to have Docker credentials for pushing.

Using the null builder

There are some occasions where you may want to run provisioners without attaching Packer to an actual builder. For example, you may want to build an image for your application using a new VM and a new container, but also connect to an existing resource and perform the same provisioner tasks used by the VM and container environments. The only thing to configure for a null builder is a communicator such as SSH or WinRM.

One particularly useful reason to use a null builder during development is that builds utilizing the null builder run the provisioners immediately without taking time to stand up an environment. If a provisioner has been configured with an error, then the null builder will usually be the first to report it, which can help you stop the build rather than wait a while for build environments to become available.

Other uses include connecting to existing environments or test environments to replicate the provisioner processes that will be used for the next image release. Note that it is up to you, as a Packer template developer, to make sure that the environment can handle repeated runs. The environment should be reset if any of your provisioners are not idempotent so that they cannot reliably be run more than once. We will cover more about provisioners in the next chapter.

Summary

In this chapter, we saw that there are a lot more integrations available to Packer than just VirtualBox. The integration between Packer and an environment or cloud to be used for building images is called a builder. In a Packer template, one or more builder is declared as a source. Then, a build job is declared with the list of sources you wish to run. Builders define a communicator such as SSH or WinRM to connect to for running provisioners. However, builders aren't of much use without provisioners to run tasks over the communicator during the build. So, in the next chapter, we will cover provisioners in more detail.

4

The Power of Provisioners

So far, we have covered the basics of Packer templates and builders with a basic provisioner. Builders allow integrations to external platforms, hypervisors, and cloud providers via API integrations. Most likely, an image will need some runtime configuration before it's saved and deployed. In this chapter, you will learn about the different types of provisioners available and what options you have to add artifacts or run tasks in a container or image. Common provisioners include local files or folders that should be uploaded to the image as well as scripts or processes that should be run as part of the building. The goal of provisioners is to minimize the work that needs to be done at runtime once an image is deployed. Classic examples of provisioners include these:

- Installing to disk from classic install media, ISO, or network boot

- Applying installation profiles, kickstarts, jumpstarts, or package groups

- Running Ansible playbooks, Chef recipes, and Puppet templates and profiles

- Installing cloud or hypervisor agents

- Enabling common services on the first boot

- Container connections and shell tasks

We'll learn how to do each of these with examples, but first, we need to cover the basics of provisioners and communicators and how they work. These are the topics covered in this chapter:

- Configuring communicators

- Injecting your config and artifacts

- Seeding a file or directory into your image once booted

- Using templates to populate configuration resources with variables

- Running a script or job across all builders

- Developing provisioners with the `null` builder

Technical requirements

You will need a machine capable of running VirtualBox or QEMU for this chapter's sample source code. We will list download links for external content. If you find a link is unavailable please check the latest on GitHub for updates or substitutes.

Configuring communicators

Each source declaration has support for a communicator option. The communicator configured with the builder tells Packer how to communicate with the temporary build environment for applying provisioners. The most common communicator is SSH, while Windows VMs typically use WinRM. Docker is an exception as Packer can connect directly to the container terminal to perform tasks without a running SSH server. A communicator usually needs credentials configured to connect, such as user and password, keys, or certificate. Take this example from our `virtualbox-iso` source:

```
communicator = "ssh"
ssh_username = "root"
ssh_password = "packer"
```

You can see that we're specifying the communicator type and the necessary credentials inline. Obviously, it's a terrible idea to have plaintext SSH credentials in a Packer template. There are a few options around this.

First of all, remember we can include credentials as variables in the template so that we specify them at build time. That at least keeps the credentials out of the template, but we will still need to specify them somehow.

The second option is to use `ssh-agent` on the Packer host. This can keep a password-protected SSH key in memory for use by your Packer process. Unfortunately, keys don't expire, so if someone gets access to an unprotected key from a Packer process or log, they could compromise your permanent credentials. This option sounds secure at first, but permanent keys can pose a problem. You may wish to use signed SSH certificates instead of keys as the certificates have a **time to live** (**TTL**) and can be configured to expire shortly after the build job finishes. You can configure your own SSH **certificate authority** (**CA**) and certificates or you can proceed to option three—the most secure option—which is to use Vault again. Vault can either secure static credentials that are accessible only to your Packer process or it can automatically generate short-lived SSH signed certificates, as in the aforementioned second option. Vault can also store static credentials for WinRM, as shown here:

```
local "ssh_cred" {
    expression = vault("secret/packer", "root_pw")
    sensitive  = true
}
```

This will do a secure lookup of the password saved securely in Vault, but it's still possible a rogue script or action could leak that password via a log. You could even use an SSH key, but that too could be potentially leaked during a build. Passwords and keys do not expire. For ultimate security, use a signed certificate that has a short-lived TTL and expires after Packer is done. That way, if your certificate is leaked in a log, it will expire and be invalid even if someone tries to use it. The CA must be configured inside your base image for the SSH server to verify your client certificate. HashiCorp Vault can be used to do this for you. There are actually quite a few options to authenticate SSH, and Packer has support for bastion hosts, proxies, and `ssh-agent`. The most common options in ascending order of security are listed as follows:

- `ssh_password = local.ssh_cred`
- `ssh_private_key_file = "/home/packeruser/.ssh/id_rsa"`
- `ssh_certificate_file = "/home/packeruser/.ssh/idcert.pem"`

For highly secure builds, always use a certificate with a short TTL. Insecure passwords and keys should only be used in development labs or as you are learning Packer. Note that `ssh-agent` can also be used to store your signed certificate in memory so that Packer doesn't also get access to any passphrase on your private key. Instead, the agent must be established before Packer runs.

There is also a datasource plugin option for Packer to look up an SSH key from an external source/ URL. This is a public community plugin and should be used with caution. As it relies on external sources for the plugin, the link is subject to change and potentially malicious code being added. First, the plugin must be defined in the Packer configuration block, including where to get the source code for it. The actual repository is `github.com/ivoronin/packer-plugin-sshkey`, but the plugin path is referenced as `sshkey`. You can see an illustration of this in the following code snippet:

```
packer {
  required_plugins {
    sshkey = {
      version = ">= 1.0.1"
      source = "github.com/ivoronin/sshkey"
    }
  }
}
data "sshkey" "example" {
  name = "packer"
  type = "rsa"
}
```

The SSH key can then be referenced using the `public_key` or `public_key_path` datasource output:

```
ssh_key_path = data.sshkey.example.public_key
```

If there is no communicator available to your build environment, you may select `null`, which is usually the default communicator when none is provided. If Packer can't find a valid communicator to apply the provisioners in your template, the build will usually fail with an error. In some cases, it may be necessary to extend the wait time for SSH or a communicator to be available, or else Packer will error with a timeout in the middle of your build. The examples in our GitHub repo accommodate this with `ssh_timeout = "30m"`.

For Windows environments, WinRM is the typical communicator. This is similar to SSH in that username and password can be specified. There is no current support for keys or certificates. The most secure way to use WinRM would be to store usernames and passwords inside Vault secrets, as follows:

```
winrm_username = "youruser"
winrm_password = "yourpassword"
```

If using Vault for your `winrm_password`, you could simply use this example instead:

```
winrm_password = vault("secret/packer", "your_winrm_password")
```

It's worth noting that the QEMU builder could theoretically make use of the emulator's terminal connection via TTY, which would behave similarly to Docker's direct communicator. Unfortunately, this communicator option doesn't exist as of this writing, but it might be a nice feature addition in the future.

Once the communicator is available in a VM, Packer will proceed to apply each of the provisioners declared in the build spec in order from top to bottom. In cases where an installation from ISO is used, there may be quite a bit of setup time required, such as kickstarting or custom installation, including one or more reboots.

Now that we've established the available communicator options, what can provisioners do with them? That's what we cover next.

Injecting your config or artifacts

Provisioners can take various forms but are limited to the abilities of the communicator. Usually, you will want to upload something to each environment. Sometimes you may want certain provisioners to be included or excluded for certain images. For example, a PowerShell script that you want to run on Windows VMs may not apply to Linux VMs or other environments. There are optional qualifiers for provisioners that can help you with complex workflows. For example, build sources may be included or excluded for each provisioner. By default, a provisioner is applied to every build source. For example, here we use the `only` qualifier to apply a provisioner to one or more sources we specify:

```
provisioner "shell" {
  only = ["qemu.base", "otherbuilder.othersource"]
```

It's also possible to override and customize the provisioner for certain sources, as in the following example. These special cases should be used sparingly, or they risk overcomplicating your build and making it hard to read. Sometimes it's better to have multiple templates than to build one large overly complex template:

```
provisioner "shell" {
  inline = ["echo default script for all other sources"]
  override = {
    base = {
      inline = ["echo custom script only for base"]
    }
  }
}
```

Another special-case option for a provisioner is to have an action run in case of errors. If your build requires a clean slate every run, it may be necessary to clean up or reset certain data in the event of a partial or failed provisioner operation. This error-cleanup-provisioner example purges a local cache directory on the local machine running Packer in the event of an error. Maybe you need to ensure a fresh download of artifacts for the next run:

```
error-cleanup-provisioner "shell-local" {
  inline = ["rm -rf /var/cache/*"]
}
```

A great example of error-cleanup-provisioner would be to download all logs from the failed environment before destroying them. You can do this with a file provisioner with direction set to download:

```
error-cleanup-provisioner "file" {
  source = "/var/log"
  destination = "/mnt/packer/ERROR_${build.PackerRunUUID}"
  direction = "download"
}
```

You can also specify a retry count to tell Packer to retry a failed provisioner a number of times if an error occurs. For example, maybe your first provisioner starts a database service or a background task but then your second provisioner will need that service to be ready, which may take a while. Try the provisioner again until it works, or error out entirely if it fails enough:

```
max_retries = 5
```

What if you want to pause before a certain provisioner to verify something before continuing or let a previous provisioner finish an asynchronous task? To do this, simply add a pause_before line to the provisioner with a duration:

```
pause_before = "10s"
```

As you customize the workflow of your provisioners, you may also have an issue where one of your provisioners is never able to finish. The entire Packer build will then hang indefinitely. There is no default timeout on provisioners. Just as communicators have optional timeouts for connecting to a resource during builds, provisioners should each set a timeout to prevent them from accidentally running forever. If there's an unusual event or a network connection becomes unavailable, a provisioner might hang indefinitely, which can become a big problem. Always set a timeout on provisioners. Here's an example of this:

```
timeout = "5m"
```

Now that we've established some common controls for provisioners, we can cover the various types of provisioners, starting with some of the most common options.

Seeding a file or directory into your image once booted

In our very first template, we demonstrated a simple provisioner of type `file` where we uploaded and replaced the `/etc/motd` file during the Packer build. This was uploaded over the SSH communicator using an SFTP operation. The `file` provisioner can actually do multiple files at once or even directories. The syntax and behavior for this are similar to `rsync`. Note that archived files will not be extracted automatically. Another provisioner can be used to extract an archive once it's been uploaded. In our first example, we set `content` to a string we wanted to write to `/etc/motd`. You may also use *herefile* syntax to upload multiple line files as content to embed a text file inside your template. It's usually a better idea to use the `source` or `sources` option to reference local files instead of embedding them. There are three main options for specifying your file content, as follows:

- `content`: Raw text to be uploaded. This does not support downloads.
- `source`: Path to a single file or directory to upload or download.
- `sources`: List of paths to upload. This does not support downloads.

We will explore line by line a sample provisioner for a file:

```
provisioner "file" {
  destination = "/tmp/"
  direction   = "upload"
  source      = "test.zip"
```

Here, we used `source` instead of `content` to specify a path to a file. This local ZIP archive will be uploaded to each build environment in the `/tmp` directory. The `source` and `content` options can't be used in the same provisioner, or Packer will error; however, if you specify `source` with a single file or directory and `sources` with a list, they will all be uploaded:

```
sources    = ["/path/file1", "/path/directory"]
```

Strings in HCL2 can also be populated with herefiles. This is helpful when multiline input is required. With a herefile, it is possible to embed a small document within HCL2 rather than include an external file. Strings can also include template variables from other parts of your code.

Using templates to populate configuration resources with variables

Here is an example of HCL's herefile or heredoc syntax, indicated by the arbitrary EOF delimiters. It may be powerful to embed complex files into one HCL file, but it can also be very hard to read later down the road. This herefile syntax is most valuable when using variables as a template. For example, here we write a multiple-line herefile to our /etc/motd file using the content option, and we include the built-in ${build.PackerRunUUID} variable so that each time someone logs in, they will be prompted with the unique build ID generated by Packer. You could also include one of your own variables or locals. Note the indentation of this syntax. The lines of a herefile in HCL are literal raw input from the beginning of each line, so if you indent your lines of text, the leading white space will be transferred to your file. This can make herefiles difficult to read in your template:

```
  content     =<<EOF
Hello world from Packer.
This is an embedded multi-line herefile.
This is build ID: ${build.PackerRunUUID}
  EOF
```

SSH is the preferred communicator for the file provisioner. WinRM is known to transfer files slowly, which means your Windows environments may need a longer timeout, or you may choose to use an SSH server in your Windows environments when dealing with large uploads. Some builders also offer HTTP or floppy disk features to share local directories into build environments without a file provisioner if needed.

Note the direction value, which defaults to upload but can also be set to download in cases where a file should be downloaded during the build. This can be helpful to download logs or events from a build environment into your Packer build machine before shutting down the instance and processing the final image.

Now that we've covered how to add important artifacts into a build over its communicator, let's cover how to do something with them. Next, we go on to cover scripts and specialized use cases where we perform actions over communicators instead of just transferring files and folders.

Running a script or job across all builders

There are several options to perform tasks in a build environment, ranging from basic scripts to configuration management tools such as Ansible, Puppet, and Chef, among others. There are also troubleshooting provisioners such as breakpoint, where you tell Packer to pause in case you need to inspect something during a build. First, let's cover the basics of the `shell` provisioner. Most tasks can generally be done with shell actions. Similar to the `file` provisioner, there are three main options to specify what you would like to run, as follows:

- `inline`: Raw list of strings you would like to run as a script
- `script`: Path to a single script you would like to run
- `scripts`: List of paths to scripts you would like to run

Take an example where we would like to update an Ubuntu environment and install an app with `apt`. This can be done inline with two lines. Remember that HCL is tolerant of hanging commas in lists such as this example. Here, we tell Packer to use our builder's communicator and run these two commands, updating the machine and installing the `cowsay` utility:

```
provisioner "shell" {
  inline = [
    "apt-get update",
    "apt-get install -y cowsay",
  ]
```

Remember—it's wise to specify a timeout to prevent a script that never ends from hanging your entire build process, like so:

```
timeout = "5m"
```

You may also want to set certain environment variables for your session—for example, HTTP_PROXY settings or other environment variables that may only be temporary for the build environment. These vary by provisioner, but in the case of `shell` provisioners, these can be specified using a couple of options, as outlined here:

- env: A map of key/value pairs

  Example: `env = { {path = "YOURPATH:/bin"} }`

- `environment_vars`: An array of key/value pairs

  Example: `environment_vars = [ path="YOURPATH:/bin" ] ]`

Either way environment variables are listed, they will be set in the process using default shell syntax (`export VAR=VALUE`). If you are running provisioners in a non-standard shell that fails on default syntax, you can override the template for these commands. For example, if using the *fish shell* (`set -x VAR "VALUE"`), you can set the `env_var_format` variable like this:

```
env_var_format = "set -x %s="%s'\n"
```

There is also a variant called `shell-local` that will run the provisioner locally on the Packer machine instead of on the build environment using the communicator. Good examples of this include connecting to a service on the build environment such as a database or a REST API. Here is an example where we have the machine running Packer use `curl` to check the build environment via an application's health REST API. If the `curl` command fails, the entire provisioner will fail, and Packer will let you know with an error. Otherwise, it will proceed to do the next command, which could be a Kubernetes port forward or a command to another service to signal success. Here, we've also used the HCL trick that strings can be herefiles, so we don't technically need to have a list of strings for each line:

```
provisioner "shell-local" {
  inline = [
    <<EOF
    curl https://${build.Host}/health || exit 1
    kubectl port-forward
    EOF
  ]
```

Elevated privileges can be obtained by piping the root password into `sudo`, though remember that having your password in your template is a major risk. If your build needs `sudo` or elevated privileges, it's wise to store the password inside Vault or—better yet—temporarily set your user's sudo access to NOPASS so that you don't need to provide the password at all during the build.

Note that when using Windows, it's best to use the `windows-shell` provisioner instead of the `shell` provisioner. Environment variables and certain parameters are treated differently between Windows and Linux/Unix environments.

The `shell-local` builder runs locally on the Packer machine instead of over the communicator. It can be used to apply external actions such as registering the build environment with agents or applying Ansible playbooks. That may sound helpful, but for tools such as Ansible, the community often makes custom provisioners specifically for these use cases:

```
provisioner "ansible" {
  playbook_file = "./playbook.yml"
}
```

Ansible is outside the scope of this book, but if you want to test, there is a generic Hello World Ansible playbook included in the repo. All it does is connect to the host via SSH and create a file in `/tmp` saying `Hello World from Ansible from Packer!`:

```
---
# playbook.yml
- name: Hello Ansible from Packer
  hosts: localhost
  tasks:
    - name: Create a file in tmp.
```

```
        copy:
          content: Hello world from ansible run via packer.
          dest: /tmp/packeransible.txt
```

Unfortunately, the plugin will only take a path to a file and not the raw playbook, so you don't have the option to inline your Ansible playbook with a heredoc. Strangely, though, you can actually still do this with a simple `shell` provisioner that calls `ansible-playbook`, although it's a messy sample of code:

```
provisioner "shell-local" {
  inline = [
    <<EOF
ansible-playbook -i localhost -c local /dev/stdin <<ANS
[YOUR PLAYBOOK HERE]
ANS
    EOF
  ]
```

This looks really ugly at first. Using nested herefiles is difficult to read. In this case, we nest a herefile marked with `ANS` inside a herefile marked with `EOF`. The good news is that you can simply embed your Packer variables right into the same file. Packer will fill in all variables before executing the provisioner. You may use similar strategies or community plugins for Puppet, Chef, and Salt. Most of these plugins provide for not just standalone runs but also server-based runs.

What about registering agents or asset managers? If building enterprise-supported images in the cloud, your updates and repositories are likely provided by the cloud. If you don't have a cloud option such as **Red Hat Update Infrastructure (RHUI)**, you may need to activate or subscribe your images during the build, but then resubscribe for each instance of the image being deployed:

```
subscription-manager register --username admin-example \
  --password secret
```

This can be done for local environments in parallel with cloud environments that don't need it. Make sure to treat any license details or passwords as secrets within your build.

Developing provisioners with the null builder

We briefly discussed the `null` builder in the previous chapter. As you develop complex provisioners, it may be useful to just test them immediately without waiting for minutes—or even hours—for build environments and communicators to be ready. The `null` builder contains only a communicator such that there is no environment to prepare. You can use this to immediately test provisioners in a stateful environment. You can even set `ssh_host` to `localhost` if you want to test the communicator using your own machine or don't have a lab machine available with SSH access. Here is an example where Packer will immediately connect to the local SSH service using `ssh-agent` so that there is no need to present keys or passwords. Packer then runs through the provisioners in order so that you can test basic functionality and syntax:

```
source "null" "sandbox" {
  ssh_host = "localhost"
  ssh_username = "youruser"
  ssh_agent_auth = true
}
```

Existing VMs or containers may use the `null` builder without any need to clone an image or deploy a new instance. This is the quickest way to iteratively develop and test your provisioners. It takes about 3 seconds to test these provisioners locally via `null`, whereas it can take several minutes for a full **Red Hat Enterprise Linux (RHEL)** VM to boot, install from ISO, reboot, and finally provision. Troubleshooting a simple error at a rate of two tests per hour will be frustrating. Make sure your provisioners work locally before adding your remote and cloud sources.

Summary

In this chapter, we covered provisioners, which are the most powerful features of Packer. The flexibility of provisioners paired with one or more builders lets you apply a little bit of customization across many different images or environments. Packer supports a powerful set of variables during the build that can be used for templating and versioning your environment's configuration. Error handling can be managed via retries and timeouts for large complex builds. Have a good look at the example code in our GitHub repo and try to experiment with provisioners yourself.

Developing the sample code for these templates rarely goes smoothly without issues. In the next chapter, we will cover tips and tricks for troubleshooting and debugging complex build jobs in a way that simplifies finding build or provisioner errors and how to fix them.

Logging and Troubleshooting

In previous chapters, we showed the basics of templates with multiple builders and provisioners providing code examples that should be free of errors. Obviously, creating these samples isn't as smooth as just writing and building. There is a lot of troubleshooting involved in most Packer journeys. As the complexity goes up, it becomes difficult to manage logs, and sometimes it's not obvious what's going wrong. Here, we'll describe some standard debugging tricks and how to manage complex logs in a parallel build. We'll also cover some tips and tricks that prepare us for future chapters, including automation and CI pipelines.

In this chapter, we'll cover the following topics:

- Managing `stderr` and `stdout`
- Using environment variables for logging and debugging
- Controlling flow and using breakpoints

First, understand that Packer uses a separate process for each build by default. Building a template with three separate AWS sources will cause them all to start at the same time. When you run a builder on the CLI, this means that all output will appear in your terminal in the order it comes. This can get very confusing when multiple sources output text at the same time. Packer colorizes each build process output to visually distinguish them, but this color isn't always preserved by automation or text viewers. For this reason, troubleshooting and keeping track of Packer logs can be tricky. Here, we'll cover a few tricks to make things easier.

Technical requirements

This chapter simply covers multiple output streams when building complex templates with multiple sources. It's important to have a terminal that supports color text. If you're reading this in black and white, some examples show color output or different outputs may be indistinguishable.

Managing stderr and stdout

Remember that Packer builds sources in parallel by forking a new process for each one and waiting for them to finish. This means the output for each build dumps its output into the same `stdout` stream at once. Unfortunately, as of this writing, there is no feature to specify one log file per process. Instead, each process output gets a random color by default. This is helpful for distinguishing build output with your eyes only if your display shows colored text. Also, note that redirecting output to a file or specifying an output file via the `PACKER_LOG_FILE` environment variable will technically preserve color highlighting, but not all text editors or tools show or preserve shell color escape sequences. In all honesty, Packer's default output is not ideal for automation, and it may be important to break output up by the process. For this, the `-machine-readable` build flag is available. Specifying this flag prefixes every line in the shell output with the process ID that created it and some metadata of where it came from. Combining the `stdout` and `stderr` streams from multiple processes can be very messy. There are a few tips that can ease reading logs after running a build, though. These are some postprocessing tricks. First, let's take a look at the two options for output: default and machine-readable.

As per Packer's documentation, the machine-readable format has the fields `0-3`, which denote the following:

- **Timestamp**: A Unix timestamp in **Coordinated Universal Time (UTC)**.
- **Target**: The artifact build name only when applicable.
- **Type**: The type of machine-readable message, usually `ui` or `artifact`.
- **Data**: Zero or more comma-separated values associated with the prior type. The exact amount and meaning of this data are type-dependent, so you must read the documentation associated with the type to understand it fully.

Have a look at our example code for this chapter, located on GitHub at `https://github.com/PacktPublishing/HashiCorp-Packer-in-Production/blob/main/Chapter05`. First, see the default raw format from building `bases.pkr.hcl`. This template includes a number of sources. Enabling two or more sources means the outputs will be color coded for each source:

```
packer build bases.pkr.hcl
```

Here's the output:

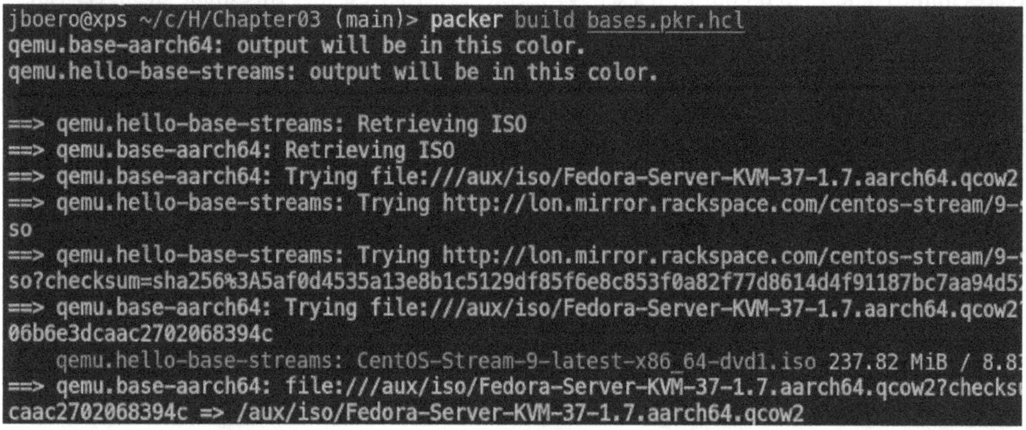

```
jboero@xps ~/c/H/Chapter03 (main)> packer build bases.pkr.hcl
qemu.base-aarch64: output will be in this color.
qemu.hello-base-streams: output will be in this color.

==> qemu.hello-base-streams: Retrieving ISO
==> qemu.base-aarch64: Retrieving ISO
==> qemu.base-aarch64: Trying file:///aux/iso/Fedora-Server-KVM-37-1.7.aarch64.qcow2
==> qemu.hello-base-streams: Trying http://lon.mirror.rackspace.com/centos-stream/9-
so
==> qemu.hello-base-streams: Trying http://lon.mirror.rackspace.com/centos-stream/9-
so?checksum=sha256%3A5af0d4535a13e8b1c5129df85f6e8c853f0a82f77d8614d4f91187bc7aa94d5
==> qemu.base-aarch64: Trying file:///aux/iso/Fedora-Server-KVM-37-1.7.aarch64.qcow2
06b6e3dcaac2702068394c
    qemu.hello-base-streams: CentOS-Stream-9-latest-x86_64-dvd1.iso 237.82 MiB / 8.8
==> qemu.base-aarch64: file:///aux/iso/Fedora-Server-KVM-37-1.7.aarch64.qcow2?checks
caac2702068394c => /aux/iso/Fedora-Server-KVM-37-1.7.aarch64.qcow2
```

Figure 5.1 – Colorized output for parallel sources

Notice that the escape sequences are shown in this raw format. Viewing this in a terminal with syntax highlighting converts these sequences— [1;32m, and so on—into colors. If you run Packer from a CI/CD pipeline or tool, you may or may not have this highlighting preserved! This is a major mistake people make when running Packer with automation. The good news is we have option number two available—packer build -machine-readable bases.pkr.hcl:

```
1656260613,,ui,say,==> null.localhost: Using SSH communicator to
connect: localhost
1656260613,,ui,say,==> null.localhost: Waiting for SSH to become
available...
1656260613,,ui,say,==> null.localhost: Connected to SSH!
1656260613,,ui,say,==> null.localhost: Running local shell script: /
tmp/packer-shell3243969037
1656260614,,ui,error,==> null.localhost: [WARNING]: provided hosts
list is empty%!(PACKER_COMMA)
1656260614,,ui,error,==> null.localhost: the implicit localhost does
not match 'all'
1656260614,,ui,message,    null.
localhost: _____
1656260614,,ui,message,    null.localhost: < PLAY [Hello Ansible from
Packer] >
1656260614,,ui,message,    null.localhost: ------------------------
---------
```

Machine-readable output includes five columns in CSV format. Instead of quoted strings, Packer replaces inline commas in output with a %! (PACKER_COMMA) escape. This single steam is easily parsed and streamed into multiple logs to keep them straight. I've created a simple script to assist:

```bash
#!/bin/bash
# John Boero
# A script to parse and split machine-readable Packer
# output into multiple files. Note this saves logs to
# arg1 or ./logs/ by default, appending along the way.

d=${1:-logs}
mkdir -p "$d"
while read l
do
    export IFS=','
    stream=$(echo $l | awk {'print $2'})
    echo "$l" >> "$d/$stream.log"
done
```

Using this script in the preceding case with two sources consisting of a null builder and a qemu builder, I can run this command:

```
packer build -machine-readable bases.pkr.hcl | ./packerouts.sh test
```

This command builds the following directory with all appropriate output in separate files:

```
test
├── error-count.log
├── null.localhost.log
├── qemu.base-aarch64.log
└── ui.log
```

Packer's default output stream in machine-readable mode ends up as ui, even though it's called from the CLI. Empty fields are skipped by the parser to our advantage. Errors are sent to error-count. log. Everything else follows the builder.name format. Now, each build log is in a separate file, which helps with troubleshooting. When using verbose builders with complex output, this is very handy.

Now that we've made sense of the stream of outputs, we can explore the more verbose options and prompt options for telling Packer to stop when we hit an error.

Using environment variables for logging and debugging

Sometimes, we need more verbose information about a build to drill down or debug more details about a build environment or failure. Packer has a few very helpful options for this. It's a bit confusing that Packer debug and log modes are separate functions. With the PACKER_LOG environment variable set to anything, Packer will dump trace output to stderr:

```
export PACKER_LOG=1
```

You may also direct it to a file instead of stderr using the PACKER_LOG_FILE environment variable. The PACKER_LOG option gives verbose information about Packer's internal operation and is compatible with the -machine-readable flag.

Packer's -debug flag shows similar output, but Packer will pause at each step of the build waiting for someone to hit *Enter* to continue. The -debug flag is helpful for manual and interactive builds but not for automated builds where nobody is present to hit a *Return* key. Unfortunately, the -debug and -machine-readable flags are incompatible, so splitting logs is not an option. It's wise to build an individual source when you need the -debug flag. Say, for example, we only want to build our QEMU example from our sample template. We can do this with the -only option:

```
packer build -debug -only 'qemu.qemu.hello-base-streams'
```

Now we have the PACKER_LOG environment variable and -debug flag sorted out, we might need some more granular control flow to step through the code more efficiently.

Controlling flow and using breakpoints

We may also need Packer to pause before continuing part of a build. Remember—by default, an error will cause Packer to destroy the environment, which unfortunately makes it hard for troubleshooting what went wrong. Rather than debugging through a build step by step, we can do some basic options to have Packer stop where we want it to.

First and foremost is the -on-error option. This can have one of four values. By default, it is set to cleanup, which explains why Packer deletes the build environment after errors. Again, as per the Packer documents, it may be helpful to set it to one of the following instead:

- abort: Exits without cleanup.
- ask: Prompts and waits for you to decide to clean up, abort, or retry the failed step.
- cleanup (default): Exits with cleanup. This is the default.
- run-cleanup-provisioner: Aborts with the error-cleanup-provisioner if one is defined.

If running Packer interactively, the `ask` option may be helpful. If running Packer from automation, there may not be an option to answer prompts. This is why `abort` and `run-cleanup-provisioner` exist. Remember—we discussed the `error-cleanup` provisioner in *Chapter 4, The Power of Provisioners*, which gives the option to retrieve logs or diagnose what went wrong before Packer deletes all build resources. Preparing for automated runs requires properly defined cleanup provisioners.

What if you programmatically want to make a build stop? Here, we can set a `breakpoint` provisioner. Don't forget we can use the same provisioner clauses—such as `only`—to limit breakpoints to a certain build or a list of builds. Note that breakpoints also require input commands to continue the build process, so they should *only* be used during interactive builds:

```
provisioner "breakpoint" {
    only = ["qemu.hello-base-streams"]
    note = "Stop to check DNS"
}
```

Provisioners run top to bottom in each build and will stop when hitting a breakpoint. Breakpoints can be disabled or commented out.

Summary

In this chapter, we covered some troubleshooting and debugging options, including how to control process flow. We also started to discuss which options are ideal for manual builds and which are better suited to automation, as we discuss using Packer with automation in the next chapter. While developing templates, it may be necessary to run interactive builds with prompts to answer questions or stop execution. For automation, you will need to make it non-interactive. In the next chapter, we'll cover how that's done.

Part 2: Managing Large Environments

In this part, we will extend the basics by combining and extending the concepts we established in *Part 1*. We will show you how to utilize multiple builders with efficiency and select a strategy for image hierarchy to save duplicate work and optimize code reuse. We will cover cloud builders and show you how to build common images across AWS, Azure, Google, and local environments. We will also explore the usage of mixed architectures and the very powerful QEMU builder to show how you can build and test your applications across cloud providers at the same time as mobile devices, IoT, and even microcontrollers.

This part has the following chapters:

- *Chapter 6, Working with Builders*
- *Chapter 7, Building an Image Hierarchy*
- *Chapter 8, Scaling Large Builds*

6

Working with Builders

In previous chapters, we showed you the basics of troubleshooting logs and output from our simple base template. Here, we will explore adding more builders to our options and also explore a feature Packer adopted from Terraform to help you look up complex attributes you might need for your build. Data sources help you look up dynamic attributes such as base image IDs or other things that may change in your builder. Rather than manually look up the AMI ID for the RHEL base, wouldn't it be much easier to search for it by name during the build? Here, we'll add builders for the major cloud service providers while focusing on some tips and tricks for each one along the way. The goal here is to prepare your image as much as possible for rapid deployment with Terraform using minimal runtime provisioners.

In this chapter, we'll cover the following topics to build on our sample template:

- Adding applications deployable from vSphere

- Adding an AWS EC2 AMI build

- Adding an Azure VM build

- Adding a Google GCP GCE VM build

- Parallel builds

- CI testing against multiple OS releases

- Pitfalls and things to avoid

Let's check the best practices for each of the major builders, including security and effectively using the data source options introduced in Packer v1.7. We'll start with the vSphere plugin and use it to cover a few basics about how builders work behind the scenes. Packer builders are currently provided by plugins written in Golang and split out into their own repositories. You don't need to be a Go expert to follow this chapter, but we will delve into plugins further in *Chapter 12, Developing Packer Plugins*.

Technical requirements

In this chapter, we do a deeper dive with builders. You will need an account with each cloud provider if you wish to test any of the related cloud examples. We also introduce some optional integration with the HashiCorp Vault product. If you wish to learn more about Vault, please check the Vault documentation pages: `https://developer.hashicorp.com/vault/docs`.

Vault is not a requirement but its integration increases the security of your Package usage and the features demonstrated are available in both the free/open source and Enterprise editions.

Adding applications deployable from vSphere

Rather than just show examples of builders, we should briefly cover how builders work. This will better prepare you to understand how data input and output work in builder plugins. The vSphere builder is actually provided by the vSphere plugin. Some HashiCorp products include all supported plugins in the product GitHub repo, while others have split plugins out into their own repositories for independent development. If you think you may have found a bug in a builder, it's tempting to go and check the base Packer source code repository at `https://github.com/hashicorp/packer`. However, if you go there and search for vSphere code, all you will find is testing and documentation code. The real bits of the plugin are stored in a plugin repository at `https://github.com/hashicorp/packer-plugin-vsphere`. This is how most plugins have been developed as products matured. This gives granular control to the community and the Alliances partners to help develop and maintain their own integrations. In this case, VMware can have full control over the plugin without full control over the Packer codebase.

Let's explore this repo to get a feel for how it presents configuration options to us and what those options match up with in the vSphere APIs. As there are multiple types of Packer Plugins, builders have a special path inside the repo. The vSphere plugins have two builders at the `builder/vsphere` path. The two subdirectories, `iso` and `clone`, provide the logic for each respective driver. Note that VMware's ISO and VMX builders for workstations are actually in a separate plugin at `https://github.com/hashicorp/packer-plugin-vmware`. The vSphere plugin is specifically used for the vSphere ISO and clone builders. There is only a `builder` directory and no data source options, or things such as post-processors. This keeps things simple to explore how builders work. Remember that a data source is a tool to look up a value in a remote system, such as a storage pool or list of images.

Each builder directory in the plugin has a `builder.go` file that establishes at a high level what the builder supports. If you browse the `builder/iso/builder.go` file, you can see an interface populated with various steps during the build process. Packer offers hooks for each step of a build. The config schema is provided via `config.hcl2spec.go`, which declares a struct and maps those struct attributes to an HCL2 schema. This declares what attributes can be defined in each vSphere ISO builder, and it's fully browsable if you need to look. It also implements the interface that describes how a vSphere build should be performed. We'll come back to this in *Chapter 12, Developing Packer*

Plugins, to write our own plugin, but for now, let's just understand where the code is if you need to troubleshoot or debug a plugin.

Most plugin repos also have very useful example directories, with sample templates demonstrating how to use them. These may not cover all use cases, but let's take a look at a sample Windows source declared in HCL2 that utilizes the vSphere ISO builder:

```
source "vsphere-iso" "example_windows" {
  CPUs              = 1
  RAM               = 4096
  RAM_reserve_all   = true
  communicator      = "winrm"
```

The source for this code is here: `https://github.com/hashicorp/packer-plugin-vsphere/blob/main/builder/vsphere/examples/windows/windows-10.pkr.hcl`.

This snippet includes just a few of the options used to declare a build of Windows 10 on vSphere. Some of these options such as CPUs and RAM are very common to any build, but let's break down the options unique to vSphere and highlight some other options not included in this example.

First, we will obviously need to configure what vCenter to use and how to connect to it. It's wise to group these together to distinguish them from the Communicator settings, which may also have a username and password. As usual, it would be wise to use Vault to secure all credentials in production. If using a development environment or self-signed certificate, `insecure_option` can be explicitly set to `true` to skip TLS verification. *Never do this in production*. Important options such as these should be set explicitly if default behavior changes in a future plugin release:

```
// vCenter connection.
vcenter_server = "vcenter.yourdomain.com"
username = "root"
password = "yourrootpassword"
insecure_connection = false
datacenter = "yourdc"
```

VMware has some unique attributes for reserved resources, such as *reservation* and *limit*, which help the build VM either efficiently utilize resources in a shared environment or maximize resource priority to build urgently. A shared cluster may also have an inconsistent performance during builds. If you find builds taking too long sometimes, it may be wise to increase the `CPU_reservation` and `RAM_reservation` values. This will require the minimum resources to be available in the vCenter used. Conversely, if the cluster is heavily loaded but time isn't an issue, then resource reservations may be omitted. Storage is configured in its own block and also may be thin- or thick-provisioned. Thick provisioning, much like resource reservations, will allocate all storage first, whereas thin provisioning will allocate it as it's consumed. If I/O performance is an issue, thick provisioning can be used, as the image is exported and compressed after builds. In most cases, thin provisioning and no reservations should be fine.

The storage block in HCL2 is actually a list of the `[]DiskConfig` custom object type, which may be unclear from the documents. This means it's specified as either one or more DiskConfig objects, which are custom-defined by the VSphere builder. `DiskConfig` is not a vSphere term but a Packer builder term. Some builders use custom objects as part of their configuration options. HCL2 doesn't actually use the word `DiskConfig` but declares instances of a `DiskConfig` object via the storage list. The best way to show this is with an example that allocates a disk of 16,384 MB, thin-provisioned in the default disk controller and storage pool:

```
storage {
   disk_size = 16384
}
```

JSON will actually require an array of objects, but in HCL2, it's possible to just specify one or more `DiskConfig` objects. A `DiskConfig` object requires a size in MB and optionally has some additional attributes, which all default to either false or zero:

- `disk_thin_provisioned`: Set to `true` for VMDK thin provisioning
- `disk_eagerly_scrub`: Set to `true` for VMDK eager scrubbing, which zeroes all storage
- `disk_controller_index`: Set to another disk controller index if the default one is not used

Here is an example storage configuration with two disks. Keep a 16 GB thick-provisioned disk for the OS and a thin-provisioned disk for the application and `/var` partitions. Remember that if you do multiple builds in parallel, thick-provisioned disks with eager scrub can dramatically slow down builds and even have a negative effect on other resources using the same storage:

```
storage {
   disk_size             = 16384
   disk_thin_provisioned = false
}
storage {
   disk_size             = 100000
   disk_thin_provisioned = true
}
```

It may be ideal to have the OS in a high-performance storage or SSD pool, but controlling this is tricky with the `DiskConfig` storage block. For this kind of control, it may be wise to use the vSphere clone builder to clone a pre-configured VM instead of the ISO builder.

Remember that Packer is about building images, but we may actually create multiple images from this build. To configure how multiple images are handled, an `export` block can be used to specify the output file in the OVF format, containing all disks:

```
export {
   name = "iso_base"
```

```
    output_directory = "./output_vsphere_iso"
}
```

One or more ISO images can be mounted as media with `iso_paths`. The first one by default will be treated as the boot image. It may be wise to include VMware tools also to install guest support with a provisioner. Paths may be relative to a datastore in brackets or a Content Library:

```
iso_paths = [
  "[datastore] ISO/winx86_64.iso",
  "[datastore] ISO/VMware Tools/10.2.0/windows.iso"
]
```

In addition to mounting existing ISO media, it may be helpful to combine files or folder data and mount the results within the build VM as a CD-ROM, without actually building an ISO. This does not require a network to fetch remote resources. The `cd_files` attribute can be used for this. Files and directories will be relative to Packer's current working directory. In this case, the CD-ROM should also be removed when exporting the template when the build finishes:

```
cd_files = ["./somedirectory/meta-data", "./somedirectory/user-data"]
cd_label = "yourlocaldata"
remove_cdrom = true
```

Remember that you can also add post-processors to builds to handle advanced tasks after the build. It turns out the vSphere plugin also adds support for two minor post-processors, *vSphere* and *vSphere Template*, which can use OVF Tool locally to extract, copy, or copy an image to other vSphere instances or clusters. These kinds of tasks are easily performed outside of Packer as part of a pipeline, but this option keeps a single source of truth in your Packer template if you choose to.

Now that we've explored more details about the vSphere builder, including some background on how plugins work via a local vSphere instance, let's move on to the major cloud providers.

Adding an AWS EC2 AMI build

If you don't have a development vSphere instance lying around, never fear. We'll cover cloud instances now, which are readily available to anyone. To develop these templates, I'm actually using sandbox environments from a third-party learning platform called Instruqt that HashiCorp uses for learning. If you need an isolated short-lived cloud account or just don't want to risk racking up a large cloud bill, there are a variety of tools and options like this, but Instruqt is preferred, as it uses HashiCorp Terraform in its backend and can give the user simultaneous cloud accounts across AWS, Azure, and GCP for a preset time. Once development templates have been perfected, they can be pushed to production or a lifecycle managed by HCP Packer, which we'll cover in *Chapter 9, Managing the Image Lifecycle*.

AWS has an extensive API to manage images and AMI libraries. For workloads that aren't ready to go cloud-native, traditional VMs and AMIs are often necessary. Remember that the move to the cloud often requires or benefits from the concept of disposable infrastructure. Rather than patching a thousand live VMs, it's common and even recommended to patch an entire image and redeploy it. This is also very beneficial with things such as autoscale groups and load-balanced clusters.

The first thing needed is a set of credentials with minimum permissions to create images. If there is no default VPC available, those permissions will also be necessary to create a VPC first. Once a VPC is available, the minimum permissions are described in the Packer documentation. The most secure Packer builds will either run on an EC2 instance, using its instance role for authentication, or use HashiCorp Vault to issue short-lived credentials for each build, with Vault using a role as a template to grant IAM privileges. Minimal Packer permissions for AWS are fairly complex and are listed as follows:

```
"ec2:AttachVolume",
"ec2:AuthorizeSecurityGroupIngress",
"ec2:CopyImage",
"ec2:CreateImage",
"ec2:CreateKeypair",
"ec2:CreateSecurityGroup",
"ec2:CreateSnapshot",
"ec2:CreateTags",
"ec2:CreateVolume",
"ec2:DeleteKeyPair",
"ec2:DeleteSecurityGroup",
"ec2:DeleteSnapshot",
"ec2:DeleteVolume",
"ec2:DeregisterImage",
"ec2:DescribeImageAttribute",
"ec2:DescribeImages",
"ec2:DescribeInstances",
"ec2:DescribeInstanceStatus",
"ec2:DescribeRegions",
"ec2:DescribeSecurityGroups",
"ec2:DescribeSnapshots",
"ec2:DescribeSubnets",
"ec2:DescribeTags",
"ec2:DescribeVolumes",
"ec2:DetachVolume",
"ec2:GetPasswordData",
"ec2:ModifyImageAttribute",
"ec2:ModifyInstanceAttribute", "ec2:ModifySnapshotAttribute",
"ec2:RegisterImage",
"ec2:RunInstances",
"ec2:StopInstances",
"ec2:TerminateInstances"
```

Let's add a builder for our gold image to be built in select regions of AWS. The Amazon plugin for Packer supports quite a few builders, as well as data sources and at least one post-processor. First, there are basic equivalents to the vSphere plugin. Amazon supports *EBS* and *instance* builder types, which you might compare to vSphere's *ISO* and *clone* builders. The first starts with installation media or a stock image and produces your own image. The second provisions a VM instance from an AMI backed by instance storage.

Amazon also offers two additional options that are advanced use cases but can save you time if you use them properly. The *chroot* builder can be very helpful at speeding up builds, as it doesn't require an EC2 compute instance to build. Instead, a volume is cloned and mounted as a chroot without booting it as a VM. The ability to customize storage without actually booting it is very handy but also very delicate, as you can easily corrupt storage or a boot option if storage changes are made without a running OS. The *EBS Surrogate* builder performs a similar operation, but you can start from scratch and build your image completely using a chroot. Gentoo users may love this one, as the traditional way to install Gentoo was always by chrooting into freshly partitioned storage. This is the concept that is applied to cloud images. Let's add a builder source for our base image to be built in AWS.

The first thing we're going to need to build a custom AMI in AWS is an existing image ID. Early versions of Packer required looking up the ID for the image you want to start from. These IDs often change as images are updated and lifecycle promoted. This is another challenge that has been simplified with the introduction of data sources. We can use the Amazon plugin's data source to look for an image ID by name and release. Let's face it – AWS does not make image discovery easy. Using the AWS CLI to search for your image can be tricky. Let's say we need the latest RHEL 9 image as a base. We need to look up the owner ID for Red Hat and properly filter those images by name, architecture, and/or region to make sure the base image is in our preferred region. Then, we need to find the right CLI query and pipe the JSON to JQ. Here, we find all IDs for Amazon Linux 2:

```
aws ec2 describe-images \
  --owners amazon\
  --region eu-west-2 \
  --filters "Name=name,Values=amzn2*x86_64*" \
   --query 'Images[*].ImageId'
```

Instead of running this before each build to find the latest RHEL 9 image, we can look it up with this simple data source in Packer. An optional `assume_role` can be declared inside Amazon data sources if necessary, but by default, the standard environment variables for AWS credentials are used:

```
data "amazon-ami" "el-amd64" {
  filters = {
    region                = "eu-west-2"
    name                  = "RHEL-9*x86_64*"
    root-device-type      = "ebs"
    virtualization-type   = "hvm"
  }
```

```
    most_recent = true
    owners      = ["309956199498"]
}
```

Now, the outputs can be consumed from this data source and used like a Packer variable, using the `data.amazon-ami.el-amd64.id` notation instead of manually hardcoding the ID every time it changes. If the image search comes up empty, it will error and stop the build from finishing. There are also more outputs available from this data source. The full list includes `id`, `name`, `creation_date`, `owner`, `owner_name`, and `tags`.

In addition to the `amazon-ami` data source, there are also sources for both a novel parameter store and a secrets manager. The AWS parameter store can be very powerful and equally hazardous. AWS Systems Manager Parameter Store can store variables and config outside of your Packer templates. This is very handy to keep Packer images in sync with your AWS config, but beware that changes outside of Packer may cause builds to fail and be difficult to troubleshoot. Declaring the data source is simple enough and the value output contains the config parameter. Each data source instance can look up a single parameter:

```
data "amazon-parameterstore" "packerparam" {
  name = "packerparam"
  with_decryption = false
}
```

Secret Manager can also be used to look up secrets in AWS Secrets Manager, similar to HashiCorp Vault. Again, each instance can retrieve a single secret, but as secrets are JSON objects, they can be parsed into HCL maps with `jsondecode`, or the data source can specify a single key to fetch in its value output:

```
data "amazon-secretsmanager" "yoursecret" {
  name = "packer_test_secret"
}
jsondecode(data.amazon-secretsmanager.yoursecret.secret_string)["key"]
```

Now that we have data sources available, it's time to create a source. We will apply our base provisioners to an RHEL 9 build in AWS and store the output AMI in an EBS volume:

```
source "amazon-ebs" "gold_rhel9_latest" {
  ami_name        = "gold_rhel9_latest"
  communicator    = "ssh"
  instance_type   = "t2.micro"
  source_ami      = data.amazon-ami.el9-amd64.id
  ssh_username    = "root"
  deprecate_at    = timeadd(timestamp(), "240h")
}
```

Here, we have a simple VM being built in the default region and VPC for our AWS credentials. The SSH communicator will be used, but we haven't provided `ssh_password`. The Packer machine may run `ssh-agent` so that Packer can automatically get its SSH key or signed certificate from the agent. By default, Packer will create a new key pair and inject it into the VM.

If cloud spend is an issue, then costs can be controlled using AWS Spot Instances. These instances are meant to be short-lived, and with a more flexible SLA than normal instances, they come at a heavy discount. Since Packer-build VMs are short-lived anyway, this is a great use case for Spot Instances. The Amazon plugin gives great control over what type of Spot Instances you'd like a build to utilize. If a spot instance fails or gets reaped during a build, then Packer will error out and let you attempt to rebuild. To utilize Spot Instances, just replace `instance_type` with `spot_instance_types` and provide one or more types of Spot Instances for the build. Packer will select the cheapest option available at build time up to the maximum of the optional value of `spot_price`.

It's very important to manage the life cycles of images and remove them frequently as new builds introduce updates and patches. An often overlooked option within the Amazon EBS builder called `deprecate_at` allows you to specify a future time for an image to automatically be deprecated so that it isn't used by accident. Here, we will build these frequently, and rather than hardcoding a date, we can specify 10 days or 240 hours from build time. We will cover more lifecycle management with HCP Packer in *Chapter 10, Using HCP Packer*.

Using HashiCorp Vault for credentials

Vault is HashiCorp's secrets management product and the number one product by enterprise revenue at the time of writing. It is a very powerful service that could fill a book twice the size of this one, but here, we will cover the basics of how it can be used for Packer. If you have never used Vault before but you are curious, I recommend you hop over to HashiCorp's Developer portal for an introduction. There, you can discover how to create a local HashiCorp Vault for development in a few seconds using the following command:

```
vault server –dev
```

Don't let the name fool you. Vault is much more than a simple secure box where you store and retrieve secrets. It essentially acts as a DHCP server for short-lived randomly generated credentials. Vault has configurable leases for the **time to live** (**TTL**) value of a credential and how long it may be renewed. For Packer, there are three example use cases you may be interested in:

- Generating a five-minute cloud credential for Packer to use

- Generating a five-minute SSH certificate for provisioner communicators

- Inserting Vault secrets into HCL templates

These five-minute TTLs are just examples but a good start. If someone sees your Packer logs after five-plus minutes, then any credentials that may have leaked are guaranteed to be invalid, as Vault will have revoked them or they will have expired. As a service, Vault runs in the background with lease timers on all active dynamic credentials, waiting to clean up expired credentials. As we are discussing builders, we will focus on cloud credentials for now. Use cases for SSH certificates and HCL templates will be covered in the next chapter when we talk about provisioners.

Using HashiCorp Vault to generate a service account for AWS is fairly straightforward. Assuming you have an initialized Vault with root or admin credentials, it just takes three commands to configure once your Vault instance is running:

1. Enable the AWS secrets engine:

    ```
    vault secrets enable aws
    ```

2. Configure service credentials that have IAM admin privileges:

    ```
    vault write aws/config/root \
        access_key=$AWS_ACCESS_KEY_ID \
        secret_key=$AWS_SECRET_ACCESS_KEY \
        region=us-east-1
    ```

3. Write one or more roles as templates for your Packer builds to fetch:

    ```
    vault write aws/roles/packer \
        credential_type=iam_user \
        policy_document=@vault_aws_packer_policy.json
    ```

At this point, Vault can be used to generate timed dynamic credentials for Packer builds with a lease TTL. If anybody in your Packer code manages to leak cloud credentials into logs, the credentials should be expired by the time logs are seen:

```
$ vault read aws/creds/packer
```

Vault will automatically revoke the credential after its TTL.

Adding an Azure build

The Azure plugin is a bit simpler than the Amazon plugin in that it currently has no data sources and fewer builder options. It resembles the Amazon plugin in that it has a main builder that uses ARM, similar to Amazon's EBS builder. Azure also supports a chroot builder, much like Amazon, that can speed up builds. Here, we'll show a sample source for each. The chroot example must also be run from an existing EC2 instance, but we'll demonstrate chroots with Azure VMs.

Packer's credentials must have certain roles configured for access to manage VM instances and generate images. Azure makes this very simple compared to other options. There are just two roles required generally – Managed Identity Operator and Virtual Machine Contributor. If you use chroot builders, you don't need to create a VM, and technically, you could have an extra-secure Packer process using just chroots without Packer ever needing privileges to manage VMs. The chroot builders thus offer the option of one less attack vector.

First, let's start with a traditional image using ARM. Credentials are looked up via the metadata API if you run Packer in Azure. Otherwise, the Azure plugin does not automatically look up the standard environment variables for authentication like Amazon, but the good news is that we have the env function in HCL2 to populate things such as `client_id` for us. env is officially reserved for input variables, which makes it a really handy default value:

```
variable "ARM_CLIENT_ID" {
  default = env("ARM_CLIENT_ID")
}
source "azure-arm" "gold_rhel9" {
  client_id       = var.ARM_CLIENT_ID
  client_secret   = var.ARM_CLIENT_SECRET
  subscription_id = var.ARM_SUBSCRIPTION_ID
  tenant_id       = var.ARM_TENANT_ID
```

Image discovery is much simpler in Azure. There is no need to look up an image ID. We simply need `image_publisher` and a few details about the image we want to use:

```
  os_type         = "Linux"
  image_publisher = "RedHat"
  image_offer     = "RHEL"
  image_sku       = "9.0"
  image_version   = "latest"
```

This simply builds our base image similar to the AWS source from publicly available base images. This also requires a default VPC to be available for the Azure user. As usual, a VM is created with our details, and an image will be built traditionally using compute resources. Remember that this can take quite a while, so let's explore the chroot option.

The tricky bit about chroot builders is that they require a running instance inside the cloud to essentially mount a copy of an image and chroot into it. As the machine will be mounting filesystems, it also requires elevated privileges, running Packer as root, or using sudo. The great part is that a build often takes seconds instead of minutes, like when generating an entire VM. It's also super simple to declare a source. This simple source assumes that Packer will be running on a VM in Azure, with permission to mount a copy of the source image given and run our provisioners. It has a source image name input and an image resource ID output. As the image is mounted locally, the OS and architecture are expected to be the same as the host running Packer:

```
source "azure-chroot" "gold_rhel9" {
  image_resource_id = "/subscriptions/YOURSUB/resourceGroups/YOURRG/
providers/Microsoft.Compute/images/gold-RHEL9"
  source            = "redhat:rhel:9:latest"
}
```

This simple block of code gives us an image with no fuss or details, and the image is created just in a few seconds using an already-running machine. Be careful when building chroot sources in parallel, as Packer will operate multiple tasks in a shared environment with other builds in the process.

Note that another advantage of chroot builds is that you don't need to have an OS in the volumes. You can technically use chroots to build additional application storage or supplemental configuration without actually installing an entire OS. Image strategy may include a single gold image for your OS and multiple options of additional storage volumes to mount into a provisioned VM, for each application or department.

Azure supports resource groups, which can keep your Packer resources contained and easy to clean up if something goes wrong. It is a good idea to create a Packer resource group.

Using a Vault to generate a service account for Azure is fairly straightforward. Assuming you have an initialized Vault with root or admin credentials, it just takes three commands to configure:

1. Enable the Azure secrets engine:

    ```
    vault secrets enable azure
    ```

2. Configure credentials that can create new service account tokens:

    ```
    vault write azure/config \
        subscription_id=$AZURE_SUBSCRIPTION_ID \
        tenant_id=$AZURE_TENANT_ID \
        client_id=$AZURE_CLIENT_ID \
        client_secret=$AZURE_CLIENT_SECRET \
        use_microsoft_graph_api=true
    ```

3. Write one or more roles as templates for your Packer builds to fetch:

    ```
    vault write azure/roles/packer \
        application_object_id=<packer_app_obj_id> \
        ttl=15m
    ```

To fetch a short-lived credential for Packer, execute this command:

```
vault read azure/creds/packer
```

Vault will automatically revoke the credential after its TTL.

Adding a Google GCP build

Google Compute is actually the simplest of the three large cloud plugins. It offers just a single unified builder that boots a VM for each build. The advantage is simplicity. The disadvantage is that there is not currently a chroot builder option in GCP to accelerate builds. Also, the Google Compute plugin makes heavy usage of the `gcloud` CLI, which must be installed on the Packer host for some use cases. Note that Google has fairly strict regular expressions on resource names, so underscores and dots are not allowed in most cases. However, GCP is happy to provide the expected regular expression when a naming issue occurs:

```
'(?:[a-z](?:[-a-z0-9]{0,61}[a-z0-9])?)'
```

When adding GCP, there is a fresh set of environment variables that are handy to import as variables. These can be defaulted whether or not they are actually present as environment variables. An unset environment variable will result in a valid NULL variable in Packer, which may or may not be an issue. If an environment variable is assigned to a required parameter, then Packer will error and require the variable to be set manually:

```
variable "GCP_PROJECT" {   default = env("GCP_PROJECT")
}
variable "CLOUDSDK_CONFIG " {
  default = env("CLOUDSDK_CONFIG ")
}
```

Credentials for the `googlecompute` builder are either manually specified if building on a workstation or obtained automatically via metadata services when running inside a GCP instance. As usual, Packer's credentials will need some permissions to manage VMs. In GCP, the minimal roles required can be tricky. The minimal privileges required for Packer are fairly simple, but in practice, other resources may need to be pre-configured, such as VPCs and storage. There are just two minimal roles, Compute Instance Admin (v1) and Service Account User. There is currently no chroot build option in GCP, so VMs must be created:

```
source "googlecompute" "gold_rhel9" {
  project_id = var.GCP_PROJECT
  source_image = "rhel-9-v20220719"
  zone = "us-west1-a"
  machine_type = "e2-small"
  preemptible = true
  instance_name = "packer-gold-rhel9"
  disk_name = "gold-rhel9"
 #account_file = "/${var.USER}/.config/gcloud/credentials"
  ssh_username = "packer"
}
```

Again, this plugin does not always use standard environment variables, so we must specify a `project_id` every time, although here, the GOOGLE_PROJECT variable is set up with the `env()` function for simplicity. If gcloud has valid default credentials set, then the `account_file` parameter is not necessary.

This OS disk from our example will be created without Google's standard encryption at rest. The disk can still be accessed by anyone with access to images in your project. If you need to control the key lifecycle, access, location, and so on for the encryption of the image, you can use a **Customer-Managed Key (CMK)** with KMS:

```
disk_encryption_key {
    kmsKeyName = "projects/${var.project}/locations/${var.region}/
keyRings/computeEngine/cryptoKeys/computeEngine/cryptoKeyVersions/4"
}
```

There are also two post-processor options for the Google plugin to import or export images after creation. These can be handy to copy, move, or archive versions of your Google Compute image, or to copy images between projects. Images can be shared between projects within an organization. It's also easy to copy images between projects manually with the `gcloud` command:

```
gcloud compute --project=destproj \
    images create gold_rhel9 --source-image=packer \
    --source-image-project=project
```

Using a dedicated project for GCP can also help contain build resources and simplify cleaning up if there is a failure or stray resources, which incur cloud costs. Packer does not have a sense of *state* as Terraform does, so it isn't easy to purge orphaned resources caused by a connection failure or another error during builds. HCP does offer some sense of state via metadata tracking, which we will discuss in *Chapter 9, Managing the Image Lifecycle*. For safety and cost savings, you may choose to use preemptible instances such as AWS Spot Instances. Preemptible VMs are well suited to Packer builds as they provide major cost savings and are designed to not live for a long time. In the rare case of an instance being preempted during a build, Packer will fail and let you know.

Using a Vault to generate a service account for GCP is fairly straightforward. Assuming you have an initialized Vault with root or admin credentials, it just takes three commands to configure:

1. Enable the GCP secrets engine:

    ```
    vault secrets enable gcp
    ```

2. Write a service account that has a role that can create new tokens. Use a credentials file with permissions to create credentials for image management, or leave this blank to use the default identity if running Packer from a GCP VM or Pod:

    ```
    vault write gcp/config credentials=@packer.json
    ```

3. Write one or more roles as templates for your Packer builds to fetch:

    ```
    vault write gcp/roleset/packer \
      project="your-packer-project" \
      secret_type="access_token"  \
    ```

```
token_scopes="https://www.googleapis.com/auth/cloud-
  platform" \
  bindings=@bindings.hcl
```

To generate a short-lived GCP credential for Packer, execute this command:

```
vault read gcp/roleset/my-token-roleset/token
```

Vault will automatically revoke the credential after its TTL.

Parallel builds

Remember that builds happen in parallel by default. This is more apparent now as we add multiple sources to our build. Packer's strength becomes more apparent in a multi-cloud deployment. Remember that Packer is a tool, not a platform or a service. Packer can be run whenever or however you want it to, but it must also be used responsibly. If you accidentally kick off 50 builds in parallel, they will all be started simultaneously. There is also no locking or state file, as with Terraform, so you may accidentally run the same build multiple times simultaneously, which will probably not be what you intended. If you use unique identifiers for each build instance, they will all build simultaneously. If you use fixed naming, then likely Packer's builders will error, saying that certain resources it needs to create already exist and can't be replaced.

In the early days of Packer's in 2017, HashiCorp built a managed platform called Atlas, which was meant to provide a service to support HashiCorp's tools. When you build a platform around a tool such as Packer, it handles workflows and pipelines for you on a first-come first-serve basis. Atlas was superseded by the **HashiCorp Cloud Platform**, also known as **HCP**. Packer is still a tool that you use as required, but HCP adds metadata and lifecycle management. Most users will not run Packer on the CLI, but they will place the tool into an automation pipeline or CI/CD platform that automates its actual runs. However, it's important to get familiar with the manual runs of Packer on the CLI before automation, which we will cover in *Chapter 11, Automating Packer Builds*. A good CI/CD strategy can help manage parallel builds and make sure that builds don't overlap in a way that breaks desired behavior.

CI testing against multiple OS releases

In this chapter, we discussed additional builder types for cloud and VMware, and we added a single example of each to our sample code. With a single-build template crossing multiple sources with multiple provisioners, we can build a single image across multiple clouds with ease. If we need to test multiple base images or OSs against our application, that is a different problem to solve. You might think of image complexity as a four-dimensional problem, described like this:

*C = (Base OSs) * (Clouds) * (Provisioners) * (Architectures)*

We have just built images for a single OS (RHEL 9) on multiple clouds (AWS, Azure, and GCP). We also assumed that each build would be running Intel's x86_64 architecture. If AWS Graviton instances go on sale with savings of 40%, your business can benefit from having ARM images ready to switch. If AliCloud suddenly releases RISC-V instances with savings of 50%, then the business could save even more by also being prepared for RISC-V. The business will save money by being prepared for tomorrow's cost savings. This is why the QEMU builder is your best friend. With QEMU, you can build and test environments locally for any platform that QEMU supports for free. Developing Packer templates in the cloud costs cloud resources and can be pricey. It's actually more cost-effective to develop templates locally first and push them to the cloud with a release. Remember that Packer itself is a cross-platform binary and can run on anything, from an enterprise server to a commodity device or ARM-based Raspberry Pi. QEMU can virtualize just about any architecture on just about any other architecture, which means you can even emulate large complex builds on a few watts of electricity using a cheap ARM device. The GitHub repository for this chapter has examples of QEMU for mixed architectures.

HCL2's dynamic code construct can be used to simplify multiple attributes within a root object, but it cannot be used to create root objects, so unfortunately, it's not possible to dynamically create sources, for example, to create five separate sources using one dynamic statement for different base images on AWS. Unfortunately, sources must be created individually and statically. If the number of sources in your template will be large and cumbersome, it may actually help to use JSON templates to create reusable sources.

If we were to add Ubuntu and Windows builds to each of these cloud environments, we would need to define multiple sources of each builder type. This can be simply done, one at a time but doing this can also complicate troubleshooting, as multiple parallel environments are built all over the world and you may really just want to troubleshoot one environment.

Remember that a Packer build has the `-only` parameter that allows you to specify the only sources you wish to build. This should help you limit which builds you want to test during development. Unfortunately, this means different things between legacy JSON and modern HCL2 templates. In HCL2, this refers to the *name* field of the source block. In the sample template for this chapter, I've used `gold_rhel9` for a common source name across all builder types to reflect the OS used across all builders. In this case, if a build is run with `-only=gold_rhel9`, then all cloud sources will be run at once. This may or may not be desired. Wildcards are supported in the `-only` flag as long as they are included in quotations. Therefore, it may be wise to combine the builder name and OS name with each source name. In that case, it's possible to test all builders against the OS by running `packer build -only='*rhel9*'`, for example. Otherwise, you can exclude builders you don't want to run with the `-exclude` option, which is the opposite of the `-only` parameter.

Pitfalls and things to avoid

A lot of example templates in GitHub and community resources only build one item at a time. It's actually more helpful to build everything you can in one template and run just the builds you need. Unfortunately, Packer doesn't combine all HCL files or all files in the current directory as Terraform does.

This means you need to put all sources and build combinations in one file, which can be problematic as templates grow. Unfortunately, you also can't include or refer to other templates, which might distribute complexity among other files. What you can do is make use of HCL2's language features.

Luckily, there are a few tools that HCL2 or JSON in Packer 1.7+ give you to simplify complex builds. Dynamic features aren't available in JSON templates, but you can actually use the HCL2 for_each construct. for_each in HCL2 isn't quite like any other foreach in a language you've used before. HCL2's for_each tends to be a dynamic code block that basically copies and pastes itself for each of the items in a list or array you give it. Instead, here, you need to prepare for what it represents via an example:

```
build {
  source "vsphere-clone.delta" {
    for_each = local.similar_builds
    vm_name = value.vm_name
  }
}
```

This tricky bit of code uses a nested value, build.source.for_each, to replicate the actual source. This essentially builds a complex set of code using a parallel array. Each element in the local similar_builds variable (which is a list/array) creates an instance source. Unfortunately, HCL2's dynamic code construct isn't very helpful in its current state.

HCL2 does not currently support any embed, include, or requires construct to share common HCL2 files, so you must include all of your code in a single file. This can often result in very large unmanageable Packer templates. Using build automation can assist with this, and it's easy to combine multiple files into one template for each run and eliminate duplicate code.

Here's an example:

```
$ cat aws.hcl az.hcl build.hcl pro_app1.hcl > combined.hcl
$ packer build combined.hcl
```

This technique can be used in situations where HCL2 can't be combined by the tool or parser, but multiple files are distinct in the access or development cycle. Another technique is to use JSON templates and combine multiple JSON documents. We will cover a bit more of this next in *Chapter 7, Building an Image Hierarchy*.

Vault also has some things to consider when integrating with Packer. Beware that Packer's vault function performs a read even during validation. This means that if you include a Vault lookup, it will be read with every packer validate instance. If you use dynamic secrets, this could potentially result in a newly generated secret with each validation, so be careful when using automation to validate templates.

Summary

This chapter explored more on builder options than the initial builder featured in our first template. This overview of builders and the plugins that contain them, introduces support for the major cloud providers. We've extended our sample template with cloud examples in a bit more depth. We have done everything so far in a single template. We also covered some tips to use HashiCorp Vault to generate short-lived credentials for Packer builds. This is very helpful for high-security environments, where logs will never contain valid credentials used in your cloud environment.

Next, we will go beyond technical validation and delve into design patterns, looking at how to make a hierarchy of base and supplemental images in an optimal structure. We'll also look further at parallel builds and the use of chroot builders to help speed up specialized image builds without creating a VM instance.

7

Building an Image Hierarchy

In the previous chapter, we covered the bulk of Packer's cloud plugins and builders and how they function in general behind the scenes. We also added some additional build sources to our single flat template. By now, it should be apparent how managing all resources through a single file can be difficult to manage, particularly when you're building application images on top of your base images in sequential order. As Packer has no state management, how can you manage which images must be rebuilt and which may be skipped? How can we minimize wasted Packer runs and maximize code reuse? These are questions we will answer in this chapter.

In this chapter, we will cover some strategies for managing the workflow and hierarchy for your images. We will start fresh with a new series of templates and focus on some sample applications instead of the builders and platforms. We will do everything manually but design a process that is well prepared for automation later. We'll also cover a few of the non-system image builder options for applications and container images that may be built independently from the OS and can usually be built in parallel. We'll also cover some tricks for bulk running builds in a way that sets us up for automated pipelines and allows separation of duty via isolated files.

In this chapter, we'll cover the following topics, starting with container image builders and preparing common sources for accepting our sample application:

- Starting with LXC/LXD container images
- Docker container image format
- Podman/buildah plugin for the OCI container image format
- Base image strategy across multiple architectures
- Aggregation and branching out multiple pipelines

First, let's delve into the differences between system images and container images. If you are used to traditional servers or VMs requiring a full OS, then you have a solid foundation for lift-and-shift image codification. Lifting and shifting VMs into VPS services or cloud VMs is just one option for cloud migration, though. The goal of a VM isn't to just be a VM – the goal is more likely to focus on the applications running on those instances. The most efficient way to run these applications is to use containers or serverless options, which involves packaging the applications without an OS. There are still VMs running a full OS that monitor things such as hardware and virtual fans that don't exist. There is no need to worry about hardware or OSs when you package your application as a self-contained container image.

Technical requirements

In this chapter, a Linux system may be required for container support and also access to `qemu-static` binaries. These may be built from source or made available on other platforms but the examples here utilize Linux packaging and tooling.

Starting with LXC/LXD images

Early container images used plain archives of an entire directory for deployment. In 2006, Google engineers started to experiment with a new kernel feature they called **process containers**. This feature was renamed **control groups** and added to the mainline Linux kernel in v2.6.24. These control groups provided a global way to limit resources for a group of processes, including CPU, memory, and storage resources, in a way that's not available to the existing mechanism of chroots. The kernel feature itself was fairly straightforward but image management turned out to be the main challenge of containers.

What if you wanted more than a single directory archive for an application container? Orchestrators added the ability to cluster containers and also the ability to distribute multiple containers in groups. This is where Google's Borg and the Kubernetes concept of a pod originates. Fun fact: the name **pod** comes from a group of whales, and the Docker logo is a whale.

For example, an application may need to include a MongoDB service and a microservice. This results in occasionally duplicating storage requirements. What happens if MongoDB and the other containers both include the same version of glibc or some common dependency? Distributing images as a simple `.tar` archive is very simple and effective but it's inefficient. Modern container images deduplicate redundant storage by identifying common layers by checksum. LXC and LXD are supported within Packer but the runtimes have been largely succeeded by Docker and Podman, which we will cover next. First, let's highlight a simple LXD container image built with Packer. We will start with a base image of Ubuntu and repackage our changes in a new image that will be published by default to the configured LXD container registry. Configuration changes should be made via standard provisioners:

```
source "lxd" "example" {
  image = "ubuntu-daily:xenial"
  output_image = "base_ubuntu"
  publish_properties = {
    description = "Trivial repackage with Packer"
  }
}
```

The great news is that Packer can package the same artifacts in a system image at the same time as a container. This enables simple parallel packaging of your applications as both system images and container images. This is what we're going to do now.

This time, we will start from scratch while focusing on an application. We will also use a directory structure rather than building everything into a single file. We will start with a simple static website hosted on NGINX from both a VM and a container. Instead of LXC/LXD, we'll start with a Docker container and a VM.

Docker container image format

Docker revolutionized container image management when it solved the container image de-duplication problem with hashed and overlayed images. During each step of a build, a checksum is generated for the image. If more than one image shares a layer or group of files, then there is no need to store them multiple times. This is a much more efficient way to store artifacts with shared dependencies compared to LXD containers.

Docker donated quite a bit of core technology to the CNCF foundation, with the container image format being part of the Container Native Initiative. A lot of traditional Docker containers you may be familiar with are now considered version 1, with version 2 extending a backward-compatible extension of version 1. Version 2 has become the standard for OCI projects, including Podman and newer releases of Docker. Note that a Podman builder is available as an external plugin that is based on the Docker plugin. Podman was built using Cgroups version 2 and OCI image formats, which have advanced features and can be run without root permissions. Docker also supports Cgroups v2 as of the v20.10 release.

The Docker builder in Packer tends to blur the lines between sources and provisioners since some allow customization of the build without provisioners. The Docker builder provides a way to customize the Dockerfile used for a source without adding provisioners. This can conflict with other sources in your template that don't have the option of customization in the builder, so use the `changes` attribute sparingly to avoid confusion. After a build, the image can either be exported as a flat archive or committed to the local image registry. For consistency and efficiency, the safest option is to commit and publish to your image registry. This will preserve the advantages of container image layering, which are negated with `.tar` archive exports of the entire image:

```
source "docker" "base_ubuntu" {
  image = "ubuntu"
  commit = true
  changes = [
    "EXPOSE 80 443",
    "CMD [\"/usr/bin/nginx\", \"-g\", \"daemon off;\"]"
  ]
  platform = "linux/amd64"
}
```

Docker has a communicator for communicating directly via the container's interface. Luckily, there is no need to use SSH within a container for building via Packer. Docker containers are highly efficient when building with Packer, but there are a lot of options available for building container images.

Why use Packer? Packer is great at combining container images and system images across multiple environments at once. The same applications and artifacts can be built in parallel across container images, local system images, and multiple architectures. Remember how the QEMU builder allowed us to add many sources with different architectures and build VM images for mixed platforms? Docker and Podman now also support multi-arch building, which uses QEMU to accomplish this.

Here, the Docker builder supports the platform directive with an optional single platform selection. If you are running Docker Desktop or a Linux distribution that supports QEMU, you can have both Docker and Podman startup or build containers using different architectures without even booting an entire para-virtualized VM. First, you will need to install the qemu-static packages for your system. For example, in RHEL or Fedora, if you want to install all the available QEMU environments, you must use the following code:

```
$ sudo dnf list 'qemu-user-static*'
Available Packages
qemu-user-static.i686
qemu-user-static.x86_64
qemu-user-static-aarch64.x86_64
qemu-user-static-alpha.x86_64
qemu-user-static-arm.x86_64
qemu-user-static-cris.x86_64
qemu-user-static-hexagon.x86_64
qemu-user-static-hppa.x86_64
qemu-user-static-m68k.x86_64
qemu-user-static-mips.x86_64
qemu-user-static-nios2.x86_64
qemu-user-static-orlk.x86_64
qemu-user-static-ppc.x86_64
qemu-user-static-riscv.x86_64
qemu-user-static-s390x.x86_64
qemu-user-static-sh4.x86_64
qemu-user-static-sparc.x86_64
qemu-user-static-x86.x86_64
qemu-user-static-xtensa.x86_64
```

Once you have installed these packages, you can pull and run any upstream container images from any of those architectures. It takes just milliseconds for Docker to start your environment if you have already pulled the image. In this case, let's try an Ubuntu aarch64 ARM container on an Intel x86_64 server:

```
$  docker run -ti --platform=linux/arm64 ubuntu /bin/bash
root@b392186ca930:/# lscpu
Architecture:            aarch64
  CPU op-mode(s):        32-bit, 64-bit
  Address sizes:         39 bits physical, 48 bits virtual
```

By using this technique, Packer can be used to centrally build and push Docker and Podman container images, as well as system images, into the cloud for upcoming architectures, such as AWS Graviton ARM and Alibaba Cloud's upcoming RISC-V release. There is no need to provision additional compute resources for any of this, and you can often use Packer's build process to help test your application against new image releases or architectures. As the platform parameter is a string and not a list, we must build one source for each platform during our build.

Podman plugin for the OCI container image format

The Podman plugin is effectively a fork of the Docker plugin, though Podman does not support the `changes` attribute or Dockerfiles. It also doesn't come shipped with Packer at the time of writing, so it must be included manually, and a Packer `init` must be performed first to download it. This time, we will need to include the `required_plugins` block inside the Packer config so that we can get the plugin during `init`. Note that this can be risky for production as upstream contributions present an opportunity for supply chain attacks. We will show you how that works in *Chapter 12, Developing Packer Plugins*. Hopefully, the Podman builder will be mainlined into Packer and supported by HashiCorp, so this won't be an issue in the future:

```
packer {
  required_plugins {
    podman = {
      version = ">=v0.1.0"
      source  = "github.com/Polpetta/podman"
    }
  }
}
```

Podman and Docker builds may take place independently or in the same template but make sure that the versions of Docker and/or Podman that are installed locally are supported by the respective plugin. Here, we have added a Podman source to mirror the Docker source:

```
source "podman" "base-example" {
    image = "ubuntu"
    commit = true
    #export_path = "image.tar"
    run_command = ["/usr/bin/nginx", "-d",
        "daemon", "off;"]
}
```

Now that we have a fresh base of container sources, we want to prepare them all to deploy a common application. In our earlier example, we continuously built everything up in a single large template. Now, things will get more complex: we need to build a base image strategy and structure.

Base image strategy

Maybe separation of duty means that different people have control over different cloud resources or container registries. Maybe some HCL files in a template directory should be read-only while others, such as variable declarations, may be edited. We need to build out a directory structure that accommodates collaboration and also helps us reuse common code. Remember that HCL doesn't have an `include` or a `require` directive to import other files or modules. You can run Packer against a single HCL file or against a directory, which will result in the combination of all HCL files included. It's not so easy to `#include common.hcl` from a shared read-only space. Luckily, we have the simple option of symbolic links for this, where you can alias a common template file into your template directory. Here is an example:

```
$ ln -s ../common/basicconfig.hcl .
```

If you look at the GitHub repository for this project, you'll see that the single HCL2 file has been refactored into a few directories. Note that any directory that includes a `main.pkr.hcl` also has a symlink to `../common.hcl`. Now, common things such as variables can be saved centrally in one file that many Packer developers can be prevented from reading or writing to using standard filesystem permissions. Note that Git will not preserve these permissions during cloning, so GitOps will only be secure if developers don't have control over read-only files during build pipelines. Eventually, automation pipelines will be used instead of running Packer manually, which will give you more control over privacy. For now, let's have a look at this directory tree:

```
$ tree
├── asset
│   ├── ks-centosStreams8.cfg
│   ├── ks-centosStreams9.cfg
│   └── vault_config.sh
```

```
├── aws_example
│   ├── main.pkr.hcl
│   └── common.pkr.hcl -> ../common.pkr.hcl
├── common.pkr.hcl
├── docker_example
│   ├── main.pkr.hcl
│   └── common.pkr.hcl -> ../common.pkr.hcl
├── lxd_example
│   ├── main.pkr.hcl
│   └── common.pkr.hcl -> ../common.pkr.hcl
└── podman_example
    ├── main.pkr.hcl
    └── common.pkr.hcl -> ../common.pkr.hcl
```

Note that, unfortunately, provisioners are not at the root level of HCL in a Packer template. They are included in the build block. This means that provisioners cannot be held in a common file as you might expect. We are allowed to declare a build referring to external sources from the root level but provisioners must be declared inline. The HCL schema used by Packer prohibits declaring provisioners at the root level. This seems limiting to Packer and would be a nice feature request. An example could include common provisioners in a file, such as our `../common.hcl`, that allow the builds within to combine specific sources with specific provisioners. For now, it's necessary to add our sample application to every template via an identical provisioner and maintain them across templates when making changes:

```
provisioner "shell" {
  inline = [
    "apt install -y nginx git",
    "cd /usr/share/nginx/html/",
    "git clone https://github.com/jboero/hashibo.git "
  ]
}
```

This provisioner is from the Ubuntu base image. It would be different based on the base image in each source. Alpine or even the native NGINX image could simplify the build, but we can use this same provisioner across Ubuntu VMs and containers.

Pay close attention to the naming of our sources. This becomes very important when running automation. Provisioners may specify which sources they apply to directly; alternatively, when running the `packer` command, we can specify wildcards. Wildcards make it much simpler to build a subset of sources. For example, if you need to rebuild all base images, all images with a certain OS or platform, or all images with a certain application, the following examples are helpful:

```
$ packer build -only="base.*"
$ packer build -except="*.ubuntu.*"
$ packer build -only="*.app"
```

In this way, we can control groups of sources built during the current process. This is why it is important to standardize on a naming scheme that suits you. Be sure to include things such as structure, platform, and application or purpose in consistent locations.

Aggregation and branching out multiple pipelines

A directory structure can be used to logically organize images that are dependent on each other. For example, a hardened enterprise Linux build with your organization's **single sign-on** (**SSO**) or SSSD configuration may be required at the base level but then multiple application builds may be built from that base image. Structuring this as a directory tree makes automation super simple. You can recursively build directories and the directories within to build an entire library of images in order. Take this example, where there is a system base image directory and two nested system orchestrator images that install Kubernetes or Nomad on top of the base image:

```
$ tree system_base/
system_base/
├── build.pkr.hcl
├── common.pkr.hcl -> ../common/common.pkr.hcl
├── system_kubelet
│   ├── build.pkr.hcl
│   └── common.pkr.hcl -> ../common.pkr.hcl
└── system_nomad
    ├── build.pkr.hcl
    └── common.pkr.hcl -> ../common.pkr.hcl
```

Each directory includes our common variables and can easily be traversed with a simple recursive bash function. Build the current directory, which is indicated by arg `$1`. If the build fails, error out. If it succeeds, touch a `build_timestamp` file and then recursively build all directories inside the current directory. Builds can check the timestamp so that they don't rebuild images when no changes have been made to the existing template:

```
function build_dir()
{
    pushd "$1"
    if [ "build.pkr.hcl" -nt "build_timestamp" ]
    then
        packer build . && touch build_timestamp
    else
        echo "build.pkr.hcl skipping: unmodified."
    fi
    find . -type d -maxdepth 1 | xargs build_dir
    popd
}
```

This requires that there is just one HCL file called `build.pkr.hcl` in each directory. This simple bit of structure and scripting is a basic start to automation. An entire stack of related images can be built at once. However, be very careful with parallelization when recursion is involved. As a single build may contain many sources, it would not be wise to build more than one directory at a time. In the next chapter, we will cover automation, which provides even simpler mechanisms for building only changed templates. Automation will optionally use the git commit hash and eliminate the guesswork around which templates have changed. HCP Packer also supports git commit hashes, which we cover later in *Chapter 10, Using HCP Packer*.

Summary

In this chapter, we branched out from having a single large Packer template to having a structured tree of images. We also added some container images to our tree for applications. We now have base system images, application system images, and container images all in one strategy. We set ourselves up for some basic automation and bulk-building images. We are still running Packer manually though and have not added automation pipelines.

In the next chapter, we will expand on automation and implement full pipelines to trigger builds based on code changes and event triggers. This is where Packer shines and becomes collaborative as teams can contribute to different parts of template code.

8

Scaling Large Builds

In the previous section, we demonstrated strategies for a structured image hierarchy, involving building shared base images and aggregate sub-images that extend the purpose of the common base image. We used a serial build script to build several image trees one at a time. The strategy is to separate these logically so that they can be built in parallel in the quickest possible time. If a minor patch is applied to our gold image, rebuilding it across AWS, Azure, and GCP one at a time will be painfully slow, and it will take a long time to learn of errors at a later stage. When building across multiple environments and complex image trees, development time becomes very important. This will set us up for automation when, in the next chapter, we streamline Packer builds via automation pipelines. In this chapter, we will take the example code from the previous chapter and logically organize it in a way that simplifies parallel builds and storage optimization for a multi-cloud strategy.

In this chapter, we will cover the following topics:

- Speeding up your builds with parallel processes
- Preventing parallel processes from causing denial of service
- Troubleshooting logs in a parallel world
- Using image compression
- Selecting a compression algorithm for Packer images
- Selecting the right storage type for the image lifecycle
- Building delta and patch strategies

We will also refactor the code to prepare it for automation. Later in *Chapter 11, Automating Packer Builds*, we will automate Packer so that builds automatically occur based on a GitOps strategy. Commits or approvals in code will automatically trigger builds and provide feedback for a collaborative framework and platform around the tool that is Packer. For now, we will continue to use manual builds.

Technical requirements

This chapter utilizes parallel processes which benefit from a large machine with high CPU and RAM resources. To realize the benefit of parallel performance requires a system with multiple CPU cores, large amounts of RAM, and high-speed storage. Your results may vary from the results in this book.

Speeding up your builds with parallel processes

If your application takes 5 minutes to build and install for each architecture but each architecture is independent, then it makes sense that independent images can be built in parallel to save time. Best practices should be able to guide logical isolation where we can maximize parallel builds. We can easily identify some guidelines for this. Divide your strategy into three layers – infrastructure, platform, and application. Other layers are up to the developer to test. Simple pattern layers are as follows:

- **Infrastructure**: The cloud/hypervisor or infrastructure used to run your platform.
- **Platform**: **Operating systems (OSs)** and golden images.
- **Application**: Images to deploy on top of your platform.

For each infrastructure, you may run a build independently. Your image should not behave differently on AWS, Azure, GCP, or on-premises hypervisor. Each infrastructure option should behave independently. This means that you can build them independently and in parallel. Even better, this will reveal to you which infrastructure is quicker and which one reports problems the earliest. A proper pipeline will stop you at the first error even when executed in parallel. This is a good thing, even if it feels frustrating, because I would rather know about my own mistake early on, rather than wait to find out if my image worked successfully on all platforms – AWS, Azure, GCP, and VMWare. The quickest layer to find a problem helps to troubleshoot first.

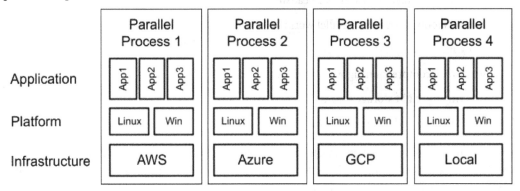

Figure 8.1 – Isolated environments for a parallel workflow

Building multiple images across public and private clouds can occur in parallel without interference. Each parallel process contains three layers. Each layer within a process may be built in parallel. Remember a Packer process will stop all parallel builds at the first error. Running multiple independent Packer processes in parallel requires the user to monitor for errors. A great strategy is to use GNU parallels to run a Packer build on each infrastructure directory in parallel. Parallel builds can fork separate Packer processes for AWS, Azure, GCP, and on-premises environments and wait for them to finish. There is a handy argument for parallel builds that will stop all processes if one of them has an error. This argument is `-halt-on-error 2`, which means you won't waste additional time and resources building other environments while an error occurs.

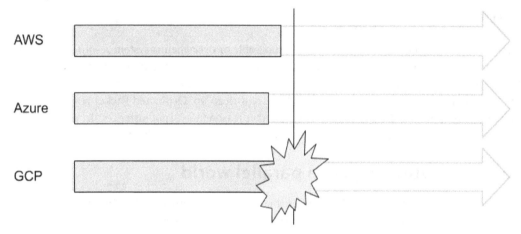

Figure 8.2 – Error handing in parallel jobs

If your private cloud or local environment is your fastest build option, it may find issues and errors even before public cloud builds, as it may take much longer. Conversely, if one build finishes completely, there is a good chance that the remaining builds should succeed too:

```
ls | parallel --halt-on-error 2 packer build {}
```

Parallel builds utilize cloud providers in an efficient way to complete images in record time, but local resources may be quickly overwhelmed by too many parallel operations. Now, we will make sure you don't accidentally cause a **Denial of Service (DoS)** on your VMWare estate or your local hypervisor.

Preventing parallel processes from causing DoS

Recursively executing parallel builds is the potentially equivalent of a fork bomb for your Packer build. On the one hand, there are a lot more cores available in today's machines. On the other hand, the cloud provides virtually unlimited threads and cores to an image builder. Each Packer template may itself contain many build sources. Parallelism and recursion together can quickly leave resources overwhelmed and unresponsive. It becomes necessary to limit the number of processes active at any given time.

We will adapt the build script from the previous chapter to use parallelism for each infrastructure grouping. Before we can automate builds in the next chapter, we also need to prepare for another potential problem — overlapping build pipelines, which must not be allowed. If a fix or change is made to code during an active build, then that build must be stopped before we can re-attempt a new build. Luckily, the GNU parallel command also supports gracefully stopping all processes. This will be very important in the next chapter when we enable code automation with CI pipelines. If a single build process fails, all other processes should stop, so we can take a look at the build logs without running any further and generating a bunch of needless logs in other processes. The simplified way to gracefully terminate all of a user's running Packer processes is this:

```
pkill -SIGHUP packer
```

At any point, if a parallel process becomes unmanageable or overwhelms system resources, leaving Packer unresponsive, then a gentle SIGHUP should gracefully shut down a Packer process, including all builds. It is not wise to SIGKILL a Packer process because there may be orphaned resources locally or in the cloud that Packer didn't get to shut down and clean up. Orphaned Packer resources will continue to cost money in the cloud! This is also another reason to use spot instances aka preemptible instances to build in the cloud. They will be terminated within a day by the cloud provider if orphaned.

Troubleshooting logs in a parallel world

We briefly touched on logs earlier. Remember that Packer randomly colorizes logs to distinguish multiple build streams appearing in the same terminal. Now that we're preparing to automate builds, we will need to plan to keep all logs for each build. They should be organized in a way that makes it easy to troubleshoot problems.

Logs should be archived and saved from each build. Store each log in its own directory, named with a timestamp or unique identifier that makes them easy to find. Retention policies should be established for archival, compression, and indexing of the logs, which can become quite large. This can be a simple time retention or a size retention policy. Using a size retention policy can help keep raw logs small enough that grep or file indexing can search the logs with ease:

- Logs less than 1 month old are kept raw in directories

- Logs 1 month to a year old are compressed as `tar.zstd` archives

- Logs older than 4 years old or your standard retention time are purged

This is simple enough to do when running Packer manually, but we will be automating quite a bit more in the next chapter via GitOps. Suddenly, a few changes of code can result in dozens of new log files being generated for each change. Some CI/CD solutions will provide this functionality for you. Others require manual work and potentially cron jobs to be sorted out. At this stage, image storage requirements are increasing, even though we're still building manual images. Before we can automate builds, we clearly need a strategy on how to store all of these large system images.

Using image compression

When building multiple images across a hybrid or multi-cloud enterprise, storage requirements can add up rapidly. In the cloud, this is apparent in storage costs, depending on how your images are stored. Global storage image buckets may be priced higher than regional storage. Different classes of storage such as SSD, standard, nearline, coldline, and archive can dramatically affect monthly cloud storage operating costs. Depending on your lifecycle and retention requirements, image storage can become a significant cost problem when storing multiple generations of images over time. Note that container images have largely minimized the problem of storage by dynamically mixing layers that are shared within a registry. Here, we are speaking solely about bootable system images.

The good news is that whether you are local or in the cloud, your images can be compressed and archived to save history while minimizing storage costs. Also, if your local system images in VMDK or QCOW2 storage use commodity storage, you may find the capital investment of local storage a far more cost-effective option for archiving purposes. Here, we will break down the strategy for when to compress and archive system images and how.

Note that Packer has a post-processor option for the compression of local images, but this actually isn't practical for some cases. Most likely, when you finish building a fresh image, you'd like to use it. Most images can't be used in compressed form. If you gzip an image, most hypervisors won't be able to use it. Instead, there are a few key steps and caveats to achieving an optimal image size and archiving it with compression after usage:

```
post-processor "compress" {
    output = "{{.BuildName}}.gz"
    compression_level = 7
}
```

This is a basic example of Packer's built-in compression post-processor. We use standard gzip with a custom compression level of 7 to give slightly smaller archives than the default level of 6. This will apply to all successful sources in the build. The strategy for this code option starts with what compression option is best for your images.

Selecting a compression algorithm for Packer images

Compression is not a simple yes or no option. There are several ways to compress an image based on the type of image and how it's formatted. Some image types support native compression and some also support encryption. The order in which encryption and compression are performed on images is important. Compression takes patterns and simplifies repetition to save space. Encryption randomizes data and makes it hard to read. If you encrypt a disk during the build and then attempt to compress it, compression will not be able to reduce the image size because the data is scrambled by encryption. Also, if your OS provides its own disk compression, then recompressing will not save any size but will take much more time to decompress before usage. Various image formats have their own standards for encryption and compression:

- **AMI**: Optionally compressed and encrypted

- **QCOW2**: Optionally compressed

- **VMDK**: Optionally compressed and encrypted via a storage device and KMIP

In the Linux world, newer filesystems commonly use compression by default. It's important to check whether your first image uses compression before deciding whether you should compress it yourself. The example code for this chapter includes a few sources to build local images from various Linux distributions locally. This allows better testing and control over the output images to find what combinations of encryption and compression work the best.

An important source image to pay attention to is the Fedora Cloud images. Since Fedora is the upstream of Red Hat Enterprise Linux and most Enterprise Linux distributions, it's a good indication of what the future has in store for things such as Amazon Linux and the cloud. At the time of writing, Fedora Cloud 36 comes pre-installed in a QCOW2 image, 436 MB in size. This is not an ISO image but comes pre-installed with a virtual BTRFS filesystem, which already uses ZSTD compression level 1. The VM will put disk I/O through compression at runtime, which negatively impacts performance but automatically gives a smaller storage footprint.

Zstandard (**ZSTD**) is a relatively new option created by Facebook for the super-quick compression of large system images. Facebook was nice enough to open source it under a dual BSD/GPLv2 license, and it has become the standard in many compression situations where speed and performance are important. ZSTD compresses and decompresses many times faster than older options such as GZIP, 7-ZIP, PKZIP, and XZ. It can also compress blocks in parallel, such that multiple cores can scale compression with almost linear performance. The workstation used to write this has to have dual Intel Xeon E5-2660 v4 CPUs, for 2x28 threads. To get a feel of how fast ZSTD is, let's take an uncompressed Windows 10 VMDK, which is 18 GB in size, and compress it with a few different options. We will use default compression levels for each of Packer's supported compression post-processor options and then do a test with ZSTD. As of the current release 1.8.3, it does not yet support ZSTD as a native compress post-processor. To use ZSTD compression, our sample code for this chapter can simply use a `shell-local` post-processor that runs our compression command directly. Use `zstd` for a single process or `pzstd` for parallel to save time:

```
post-processor "shell-local" {
  inline = [ "pzstd {YOURIMAGE}" ]
}
```

Performance metrics for our tests can be found here. All tests are run from memory in TMPFS so that no disk I/O is a factor. In practice, compression will be limited to storage speed and latency. This uses single-threaded compression for a direct comparison. Both `gzip` and `zstd` compressors manage to compress 18 GB down to around 8.8 GB, but `zstd` is about eight times faster:

```
$ time gzip win10.vmdk
18GB to 8.8GB in 762 seconds
$ time zstd win10.vmdk
18GB to 8.6GB in 95 seconds
```

Remember that this is just single-threaded. Both `zstd` and `gzip` support parallel compression. Repeating the same test with parallel compression on 56 logical CPUs gives the same compressed size but with much better performance for both. `zstd` is still far faster than the older compression standards of `gzip`, `pkzip`, and `7zip`. Compression speed isn't quite linear with a CPU count, largely because memory speed becomes a bottleneck and limits CPU utilization. Factors such as memory generation and NUMA settings may affect compression speed, based on the hardware used for builds. If you're building images in the cloud, these factors may not be visible to you. Cloud hypervisors often hide performance data such as socket topology and CPU steal metrics from the end user, so you can't see how the hardware is shared with other resources:

```
$ time pigz win10.vmdk
18GB to 8.8GB in 27 seconds
$ time pzstd win10.vmdk
18GB to 8.6GB in 9 seconds
```

Performance results are clear. Zstandard is designed by Facebook for fast image compression, and it is the clear choice where it is available. Some image formats or storage platforms may not support `zstd` yet, but legacy compressions such as gzip and `pkzip` have wider support if needed.

Packer's built-in support for post-processor compression supports the following formats. TAR is an uncompressed file that can combine multiple files and directories into a single file, and it is sometimes combined with compression to compress directory archives when directories are not normally supported by the compression selected:

- `.zip`: Easily compatible with Windows

- `.gz`: A file for Linux and the cloud

- `.tar` or `.tar.gz`: A directory for Linux and cloud

- `.lz4` or `.tar.lz4`: A fast directory compression for Linux and cloud

Most compression options in Packer give control of the compression level but not necessarily parallelism. They use the built-in Go implementations for compression, not external binaries. You may find it much more powerful to do your own compression using a `shell-local` post-processor, although if you plan to immediately use the image, it may be best not to use compression at all until an image is deprecated or archived.

The general suggestion for Packer image compression is to use parallel `zstd` in a post-processor. It will provide the best speed and compression quality while maximizing the CPU count available in your environment. Note that if building multiple images in parallel, then parallel compression options may cause a stampeding herd and potentially cause a DoS on your machine. This problem is more apparent when using automation and build pipelines. We will cover how to minimize the exposure and risk of this problem in *Chapter 11, Automating Packer Builds.*

Remember that image compression will not help if the OS installed already compresses the storage directly. A wise strategy is to select which images can use compression within the OS and which images will be raw but compressed after the build. Runtime performance will most likely be better if the OS does not compress your data for you. Databases and applications that have a lot of random I/O will perform poorly with runtime compression. Also, note that images should be properly pruned during the build. The sample code for this chapter does a few things during the build that are important for a lean and small image:

- Purge any temp files or cache
- Clean all package downloads via yum, dnf, apt, or chocolatey
- fstrim the filesystem to discard any deleted blocks
- Compact any storage layer, which will *defrag* unused blocks

When testing compression methods, we have been cheating a bit. We used tmpfs to eliminate I/O bottlenecks as a factor. The workstation used to write this has 512 GB of RAM, which can fit multiple system images in memory at the same time. Most environments will have performance issues with large storage requirements for builds. The type of storage you use for builds can make a significant difference to build speed.

Selecting the right storage type for the image lifecycle

Packer builds benefit from the fastest storage available, but once built, your large images may be compressed or archived to slower storage for cost savings. Eventually, images may be deprecated and archived in cold storage or discarded. Ideally, Packer templates are version-controlled over history. Compliance may not require you to store full images for history. Often, it's easy enough to just take a prior version of the Packer template and rebuild a historic image. This can save a large amount of space as long as the original artifacts used to build old images are still available. As long as all external resources are cached locally, it should be possible to rebuild images historically. Some retention policies may require a full history of images archived. System images may be compressed and written to cheaper classes of storage. Tape backup may seem like legacy, but it is still relevant in quite a few data centers. LTO tape still provides the lowest cost long-term archival storage in most situations, and modern compression can often fit even more content into a low-budget cartridge. In general, the following local storage classes are listed in descending order of preference, from building to decommissioning an image:

1. TMPFS (RAM)
2. NVMe SSD
3. SATA/SAS SSD
4. SAS/RAID
5. SATA
6. Tape

Cloud options are generally simpler. Some clouds have automatic class migration, which will be managed dynamically between tiers. This can be powerful for application data, but image data should be simple enough to manage on a schedule. In descending order of cost, each option costs about half of the option above it per month. Transfer fees may apply to cheaper storage, depending on the cloud and region used:

- **Standard** for data you may need immediately. This is full price.
- **Nearline** for data needed less often or in certain regions.
- **Coldline** for data on cheaper slower storage.
- **Archive/glacier** for the slowest, cheapest data kept for at least a year.

Active golden images should always be available in the quickest storage. For best performance, they should usually not be compressed in any way. Sometimes, the OS disk doesn't need high performance, but if capacity is an issue, then runtime compression should be considered at the cost of performance. Luckily, zstd de-compression is also generations quicker than previous compression methods. Decompressing the same archives created for our test reveals 72 seconds for gzip and only 10 for zstd. We use parallelization for both options, but decompressing images is usually difficult to parallelize because compressed blocks have variable lengths. zstd still managed to optimize this quite a bit:

```
$ time unpigz win10.vmdk
8.8GB to 18GB in 72 seconds
$ time pzstd -d win10.vmdk
8.6GB to 18GB in 10 seconds
```

Most specialized images aren't quite this big, but consider that decompressing a compressed image to deploy it may add more than a few seconds to your deployment time.

Building delta and patch strategies

Diffs and patches can be used between minor changes in system images, which could dramatically reduce the size of image archives. Encrypted images will likely not benefit from patches. Container images already benefit from this technique, with layers comprising an entire image and deduplicated common layers being identified by checksum. It's simple enough to generate a patch between two images:

```
diff -u oldimage.qcow newimage.qcow > 1-patch.diff
```

The patch should be much smaller than storing the entire new image if there are only minor changes and both images are not compressed or encrypted. Generating patches may be a helpful way to archive old images but is probably not practical for active golden images.

Summary

In this chapter, we discussed important strategies to consider when scaling Packer builds to large and complex image strategies. We built some simple tooling to prepare for automation later. The ultimate goal of Packer in a regulated environment is not to run builds manually but to have a collaborative workflow using GitOps, letting automation pipelines take care of the rest, which an entire team can observe. We also studied storage strategies and compression options for images when storage requirements grow. This will be very important in the next chapter when a single code commit may trigger multiple automation pipelines and image outputs. These topics will prepare you to use efficient image strategies and also ensure you don't run out of storage space with redundant or inefficient artifacts and images. The next chapter applies what we learned here in a much better workflow, and selecting the right strategy for code structure and parallel operations will become more obvious. We will take the build scripts we created in *Chapter 7* and *Chapter 8* and build an automation platform around them, with full support for starts, stops, error handling, and log analysis. There are multiple solutions in the CI/CD pipeline stage, but we will focus on GitLab.

Part 3: Advanced Customized Packer

Longer-term strategy for Packer includes lifecycle management for decommissioning old images and the optional usage of the **HashiCorp Cloud Platform** (**HCP**) Packer SaaS platform from HashiCorp. In this part, we will discuss image retention strategies and image ancestry to extend the image hierarchy concepts used in *Part 2*. We will discuss both the free tier of HCP and the paid version for advanced features. We will put together every manual step we explored earlier and automate them with continuous integration pipelines.

Finally, we will explore using Golang to write a simple plugin for `systemd-nspawn` containers. This is the most advanced topic covered in this book for when you find a new use case that Packer doesn't currently support. Previous knowledge of Go will be important in this final chapter.

This part has the following chapters:

- *Chapter 9, Managing the Image Lifecycle*
- *Chapter 10, Using HCP Packer*
- *Chapter 11, Automating Packer Builds*
- *Chapter 12, Developing Packer Plugins*

9
Managing the Image Lifecycle

In the previous chapter, we covered compression and strategies for minimizing image storage requirements and producing small lean images. Remember that compression, image layering, and container images can be used to save costs on storage for large image libraries. Storing container images as raw archives is highly inefficient and negates the deduplication advantage of container image layers. At some point, it makes the most sense to retire and decommission images altogether. Implementing a standard lifecycle management strategy can save storage waste with older unused image builds and also prevent old unpatched images and artifacts from being deployed in your environments, which might pose security risks. The final stage of lifecycle management is often the most important and involves ensuring no existing systems continue to use decommissioned images.

First, we need to establish some basics of image lifecycle management.

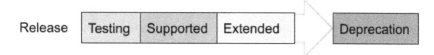

Figure 9.1 – Image lifecycle from release to deprecation

The lifecycle of Enterprise OS support often seems to start with hardened releases, but most such software also offers beta or further upstream releases. The most agile organizations will build their applications using both official releases of operating systems and platforms, and also build with upstream or beta releases to prepare for future releases and deprecation of legacy images. Efficiently maintaining images for everything from pre-release platforms through extended support OS and archival of decommissioned images constitute a well-rounded lifecycle strategy. There is no reason your application should not be ready to release at the same time as a major OS release that may contain critical updates for security and stability.

In this chapter, we cover the following topics:

- Tracking the image lifecycle
- Using the manifest post-processor
- Creating a retention policy

By the end of this chapter, you should be able to quickly design and implement an image lifecycle policy, including what metadata you may need to track and how to safely implement the purging of old images in a way that they can be rebuilt from source archives in version control.

Technical requirements

In this section, it is important to have some knowledge about industry standard compliance. Particularly, we use PCI-DSS as a standard used by payment processors. Details aren't important but reading through a full profile would be helpful for those unfamiliar with PCI-DSS security hardening. Also, JSON is used, and examples are built with the `jq` command, which should be installed.

Tracking image lifecycle

If you were to make a list of every image we've built with Packer up until now, what data would you want to retain and for how long? This criterion may depend on some regulatory requirements in your business, or it may be an internal decision for security and best practice. You may need to record only what image names are used and where they are available, including cloud storage regions, local storage pools, image archives, or container registries. Other important attributes may include system packaging metadata, such as which versions of key libraries are installed, or maybe even the results of vulnerability scans, Open Policy Agent profiles, or OpenSCAP scans. Having these on file for your entire image library can be very helpful when important CVE announcements are made, or zero-day vulnerabilities go public. Knowing which images are affected becomes very important, and there are many vendors offering solutions purely to help identify these issues. The good news is that Packer's built-in mechanisms provide some basic but powerful options already. It may be a wise decision to locally store a folder or database of output for every build you do, including things like the following:

- A full list of installed `yum` or `apt` packages for Linux
- A full list of MSI or installed Windows packages via Chocolatey
- A full environment OpenSCAP scan for your desired profile
- A list of container image packages or layer checksums
- A list of software firewall rules, if any

Let's take some public examples for inspiration. A great source of lifecycle data is provided by major Linux distributions at `https://www.distrowatch.com`, including historical releases, versions of the kernel included, and a list of important base dynamic libraries such as glibc, libc++, display managers, and more. These scans are independent of each other and thus can be performed in parallel provisioners so long as they are the last provisioners at the bottom of our build list after all configuration occurs.

First, let's kick off an OpenSCAP compliance scan against the standard PCI DSS profile and save the output report for each build. While that scan occurs, we can also run other simple scans in parallel, such as dumping a simple archive of all installed packages. OpenSCAP is free and simple with support for Linux and Windows and offers support for scans locally, over SSH, libvirt VMs, or Podman and Docker containers. The least intrusive way to scan with Packer is via a local provisioner that uses `oscap-ssh` remotely to your temporary build host. This will scan over SSH and store all output in the local machine without requiring `oscap` to be installed in the temporary build environment. The resulting scan can be viewed in a browser or SCAP Workbench display, as shown in the following screenshot:

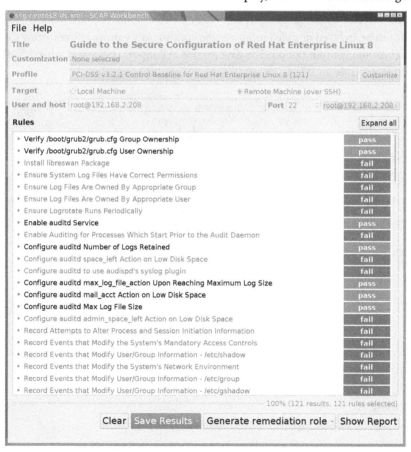

Figure 9.2 – SCAP Workbench scans against a PCI-DSS profile

The same scan can be automated with a shell provisioner or a `shell-local` provisioner. This provides useful information for archiving and should be performed before major image updates are released to identify critical system weaknesses early:

```
provisioner "shell-local" {
   inline = [<<EOF
   oscap-ssh ${build.User}@${build.Host} ${build.Port} xccdf eval \
  --profile xccdf_org.ssgproject.content_profile_pci-dss \
  --results-arf arf.xml \
  --report report.html \
  /usr/share/xml/scap/ssg/content/ssg-fedora-ds.xml
   EOF
   ]
}
```

Resulting scans can be archived or diffed using version control. Any worthwhile reporting tool should have an easily consumed output option. The data for a SCAP report is in standardized XML format with the option to generate an HTML report like the one in this screenshot:

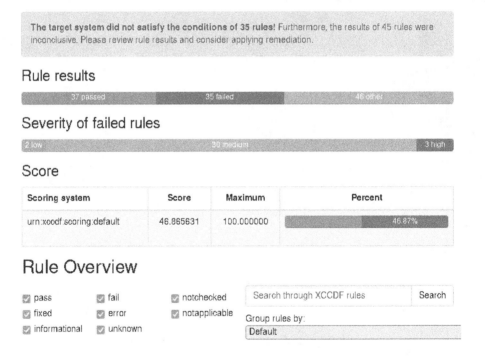

Figure 9.3 – Standard installation profile warnings

These scans can add a lot of time to a build, but they should be saved as output with every production release if you need to prove compliance during an audit. Ultra-strict environments can even use a scan to prevent non-compliant code from being released in Packer images. Also, many third-party solutions can be used for vulnerability scanning as part of your automation pipelines. Popular options include the following:

- Nexus
- Snyk
- Sonatype
- Sumo Logic
- Sysdig
- Twistlock

Commercial scanning solutions are out of scope here, but these solutions typically scan all code and libraries in an environment for known vulnerabilities and can prevent you from deploying them into production image libraries. Some can also be used at runtime to continuously monitor for defects or issues.

Whatever data you collect, it's wise to have a directory for each build and an index of what contents the directory holds. This helps track build output for the audit trail. The manifest post-processor helps to provide an index of a build's metadata.

Using the manifest post-processor

Collecting a lot of build data together can result in a mess of directories and artifacts without any way to index or organize them. Packer does provide a simple mechanism to index this data. The manifest post-processor can be used to generate a helpful JSON document listing what was generated during the build. The manifest isn't enabled by default but can be enabled by simply adding a post-processor for the manifest in just one line:

```
post-processor "manifest" {}
```

If you simply enable the manifest post-processor, Packer will use default settings, which are pretty useful, without customizing anything. This means that a JSON document will be created or appended at the `./packer-manifest.json` path. A list of all builds will be maintained within the manifest. It would be wise to either rotate this file or store it inside a JSON document database such as MongoDB if you wish to search for it later. Or, we can see in the next chapter that the HashiCorp Cloud Platform can be used to store it centrally and consumed from Terraform.

The output record for a build encapsulates metadata about the build, such as files output, timestamps, and more, depending on the plugin used. An example is broken down as follows:

```
"name": "base_ubuntu"
```

The name of the build isn't directly taken from the source name attribute but from the post-processor `packer_build_name` value, which defaults to the source name. In testing this, I found a tricky issue when using reference sources. The syntax is handled differently if you declare a root source and create a reference of it inside the build block. The root source name is used instead of any reference name. This is confusing because all sources will show the same name in your manifest. There is a way to avoid this by setting the manifest post-processor `packer_build_name = source.name` parameter, which is demonstrated in the sample repository. Builder type is simply the builder type used for this source. We use Docker for this example. Note that the manifest does not necessarily list advanced features of the builder, for example, what Docker registry we may have pushed to in a post-processor:

```
"builder_type": "docker",
```

Next is a UNIX timestamp in epoch seconds reflecting the time this single build source finished. To convert a UNIX timestamp to a human-readable date, you can use `date -d @1671299177`, which translates to `Sat Dec 17 05:46:17 PM GMT 2022`:

```
"build_time": 1670883707,
```

All of the files generated according to the source are included as a list of maps. Attributes include the filename and file size. This does not include any compression or movement after the build's manifest:

```
"files": [{"name": "riscv64.tar", "size": 64113664}],
```

The artifact ID actually reflects the type of artifact generated. Packer Run UUID is a unique global identifier for the entire run. The terminology between Packer runs, builds, builders, and sources can be confusing. There is no clear answer for this as Packer has evolved from a build that includes builders to a build that includes instances of builders called sources. A Packer Run is implicitly a single invocation of the Packer binary. All builds will be combined into a single build also known as a run:

```
"artifact_id": "Container",
"packer_run_uuid": "4555a0ff-7cc2-0f59-d685-[TRUNCATED]",
```

Custom data is actually a custom parameter so you can include custom outputs using a simple HCL2 map. In this case, we actually use a custom data source to look up the public IP of the machine running Packer. That data source is output as a custom field in the manifest. It may be handy to see what public IP was used to fetch artifacts for a build. Note that public IPs may be a security risk so should be kept protected. The example here is obviously substituted:

```
"custom_data": {"public_ip": "1.1.1.1\n"}
```

If you use the same manifest file for many builds, it can become very large and performance will be affected. It may be wise to rotate this file or store the JSON in an external JSON database, or you may use jq to extract relevant facts into an archive. Some basic jq commands can help query a large complex manifest with ease. Here are some examples. Note that records will be listed in chronological order, so there should be no need to sort by build_time:

1. List all build files. Note that they will be overwritten for each build ifthey are using the same name:

    ```
    $ jq .builds[].files[] packer-manifest.json
    {
      "name": "x86_64.tar",
      "size": 80321024
    }{
      "name": "aarch64.tar",
      "size": 71772160
    }{
      "name": "riscv64.tar",
      "size": 64113664
    }
    ```

2. Convert each build item into a CSV with build_id, build_time, and name:

    ```
    $ jq -r ".builds[] | [.packer_run_uuid, .name, .build_time] | @
    csv" packer-manifest.json
    "606e8c6b-3656-4420-d2c0-8303b5827cc9","base_ubuntu",1671368174
    "e3c3676d-6631-cd53-7a5c-11a2e87241f5","base_ubuntu",1671368811
    ```

3. List every file output and its size in CSV format:

    ```
    $ jq -r ".builds[].files[]| [.name, .size] | @csv" packer-
    manifest.json
    "x86_64.tar",80321024
    "aarch64.tar",71772160
    "riscv64.tar",64113664
    ```

Generating these queries can be useful but generally, a more robust database will be necessary for practical use. Or better yet, in the next chapter, we will use HCP, which will store and manage all metadata for us without needing the manifest post-processor. The HCP option has the added benefit of automatic lifecycle management.

Creating a retention policy

Now that we have some ways to track image metadata with each build, the next step is to design a retention policy. Retention times largely depend on compliance standards for your industry, but it never hurts to have an overprotective policy. Some Financial Conduct Authority guidelines say platform records should be kept for at least five years. Some more strict organizations may require seven years. Say we need a maximum retention of seven years for all image data. We determine that images must be retained for seven years but they can be compressed and archived after two years. We can establish these guidelines in a simple standard:

1. At three months of age, an image shall be deprecated for new deployments.

2. At one year of age, all deployed instances of an image should be terminated.

3. At two years of age, an image shall be compressed and archived in cold storage.

4. At seven years of age, an image can finally be purged.

Figure 9.4 – Image retention timeline

There is one important distinction to worry about. Immutable infrastructure usually implies that system images are not patched or maintained. Instead, they are rebuilt and replaced. This conflicts with many requirements to patch critical vulnerabilities as soon as possible. In an environment where images can't be deployed with enough speed to address critical patches, patch management can add complexity to testing and image maintenance. Forensics has better chances of inspecting a static image that was not patched when something goes wrong. Also, archived images are more useful when reproducing prior environments when there are no changes to the images. These are the major caveats when using immutable infrastructure without patch management. Ideally, immutable infrastructure can be adopted fully to quickly replace patched images as system maintenance requires.

Summary

In this chapter, we explored some tools that can be used for image lifecycle management and complex builds. The manifest post-processor can be used to capture metadata locally in a simple JSON document. The tools used here are simple and have no outside requirements or dependencies besides your own environment. The sample code repository for this chapter extends previous chapters with examples that now can export build manifests and also system image provisioners to enable dumping system profiles or system scan outputs. It is up to the user to build a workflow or way to consume the output of Packer, which, importantly, includes the manifest output in JSON format.

In the next chapter, we will use HCP Packer for this. HCP Packer serves as a central database for collaborative Packer metadata storage and automatic lifecycle management.

10
Using HCP Packer

> **Note**
> HCP is under active development and features may change frequently.

In the previous chapter, we talked about image lifecycle management, establishing generic policies for retention using the manifest post-processor, and collecting outputs for each build in a directory. The good news is that HashiCorp Cloud Platform can actually do a lot of this lifecycle management automatically. HashiCorp Cloud Platform is the only paid and supported product under Packer. There is a Free tier that can be used for up to 10 images, but advanced features of the Standard and Plus tiers cost a fee based on consumption. As a relatively new feature, HCP Packer is largely HCL-only and is not supported by legacy JSON templates. One of the biggest challenges of Packer has always been image discovery. Now that Packer has built multiple versions of your images around the globe in multiple cloud providers, how can you deploy those images with something such as Terraform? Terraform can search for all of your images around the globe, but that can be unreliable and also insecure if someone else injects images into your image registries. HCP Packer simplifies the handover from building images with Packer to locating and deploying them with Terraform code.

HCP is a global multitenant platform built to host managed instances of HashiCorp service products in one or more cloud environments. Packer is a tool, not a service. It does not run around the clock waiting for requests like many HashiCorp products. Unlike Vagrant Cloud, which stores image data, HCP Packer stores only metadata. What HCP does for Packer is it archives image metadata from every build you do. Packer is a producer of this metadata and Terraform is a consumer of it. Terraform can use the HCP provider to look up your most current golden images, and Packer will automatically retire them based on a policy, which we defined in the last chapter. This prevents someone in your organization from deploying an old or unsupported image by mistake. First, we will need to familiarize ourselves with HCP and the API used to access it.

In this chapter, we will cover the following topics:

- Creating an HCP organization
- Configuring HCP Packer in your templates
- Consuming HCP Packer from Terraform
- Using HCP image ancestry
- Exploring the HCP REST API

By the end of this chapter, you should be able to apply an image lifecycle policy using HCP Packer. This chapter will focus on cloud environments instead of local builds. We will use HCP to maintain the lifecycle of complex images across multiple regions. First, though, we need to get a basic understanding of HCP.

Technical requirements

In this chapter, we will make use of the HCP SaaS platform, which will require a free account to follow the examples. We will discuss paid features also, which can be tested with a free trial from HashiCorp if you don't have the budget for paid support. The sample code for this chapter also uses features recently added to Packer, so make sure to use the latest versions of Packer and also the latest AWS plugin. If an older version of the plugins is cached in your home directory, it may be wise to move or delete any old versions from `$PACKER_HOME_DIR/.packer.d`, which defaults to `$HOME/.packer.d`. Otherwise, we will show how to upgrade providers with `packer init -upgrade`.

The sample code for this chapter uses the Amazon plugin, which can be browsed at the GitHub repo here: `https://github.com/hashicorp/packer-plugin-amazon`.

Documentation for HCP Packer and the API are located here. Make sure to check the latest documentation: `https://developer.hashicorp.com/hcp/api-docs/packer`.

Don't forget the latest example code from our repository can be found here: `https://github.com/PacktPublishing/HashiCorp-Packer-in-Production/tree/main/Chapter10`.

Creating an HCP organization

The heart of an HCP structure is an organization. Remember that HCP is multitenant and allows a user to join multiple organizations for separate teams or business units. An organization encapsulates team-role-based access control, billing, and service groups. An organization can create one or more projects for a team to share, including multiple instances of Consul or Vault services. Each project may only have one instance of HCP Packer. Organizations and projects can be created manually via the web page: `https://cloud.hashicorp.com/`. The menu on the left of the HCP portal will show your service navigation as follows:

Figure 10.1 – The HCP navigation bar

Traditional services such as Vault and Consul can be deployed in multiple instances per project and are often configured via a Terraform provider. Currently, Packer must be enabled manually with a global registry for the project. At the time of writing, there are three tiers of HCP Packer – Free, Standard, and Plus. Be sure to check HashiCorp's Products website for the latest information, as features are subject to change. Paid Standard and Plus tiers offer support and extra features, as shown in *Figure 10.2*. This is a subset, but the full feature list should be available on the HashiCorp Products website: `https://www.hashicorp.com/products/packer`.

	HCP Free Get started	HCP Standard Get started	HCP Plus Get started
Image compliance checks (data source)	✓	✓	✓
Image compliance checks (resource)	⊗	⊗	✓
Scheduled revocation workflow	⊗	⊗	✓

Support

	HCP Free	HCP Standard	HCP Plus
Community	✓	✓	✓
Premium support and services	⊗	✓	✓

Figure 10.2 – The core differences between HCP tiers

The Free tier offers basic functionality for up to 10 images. It was also recently upgraded to support full ancestry and manual revocation, which were previously paid features. This means you can structure images together and manage the lifecycle of ancestry trees. It will not support compliance checks or scheduled revocation, which are still paid features at the time of writing. The Free tier should be enough to do most of the exercises in this chapter. Image hierarchy is managed in HCP through a feature called **ancestry**. Images that serve as base images for other images are tracked in a hierarchy and can be automatically phased out and retired from image lookups. Remember that Packer itself is not a service and does not run in the background, but HCP Packer as a platform runs background processes to expire and revoke images on a schedule. HCP Packer will only revoke metadata. It can't purge image storage at the time of writing. Metadata is universal and can be captured for just about any builder, including VMware and container images. Purging the actual image storage of retired images is up to the user.

Once an organization is created and the HCP Packer registry is enabled, you may wish to set up users. Users and Service principals tokens are managed via the IAM tab in the navigation, as shown in the following screenshot:

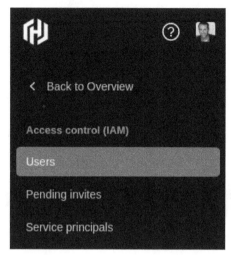

Figure 10.3 – HCP IAM user management

Users may be authenticated via a GitHub ID or by creating a local HCP account. Organizations may choose to enable single sign-on also via SAML at the organizational level. Users are global to HCP, but they may belong to multiple organizations with a different privilege level inside each organization. This RBAC principle also applies to API tokens, which will be required to use HCP Packer inside your template. There are three roles as options within HCP:

- **Viewer**: Read-only access to resources

- **Contributor**: Viewer access plus create and edit access to resources

- **Admin**: Full contributor access plus administration over users and roles

Remember that these roles cover all of HCP, including other HashiCorp-managed instances of Consul, Vault, Boundary, and Waypoint. In this case, it is wise to add a Contributor user for your Packer build to upload metadata. To consume HCP Packer via Terraform, the minimum privilege to look up Packer resources would be a Viewer role.

We will actually need to create a service principal for the role with which we want Packer to access HCP. Packer will need to write metadata to HCP but not change the organization settings, so we create a service principal called `Packer` with the Contributor role. Then, we can create a key that we will need to configure as environment variables for the API and Packer processes. This can be configured via the Packer Terraform provider or the UI, as shown in *Figure 10.5*. The keys have been partially obfuscated from this screenshot:

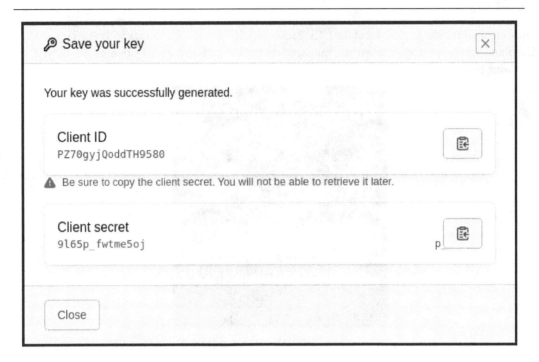

Figure 10.4 – The service principal key setup (censored)

Now that an organization and default organization with at least one user are configured, we can take a look at how to add support to a Packer template.

Configuring HCP Packer in your templates

HCP Packer support is actually implemented within the plugin repositories. For the sample code in this chapter, we will use the Amazon plugin to build a family of EC2 AMIs. Be sure you have the latest versions of the plugin available, as not all features are supported by older releases. Updating your local Packer plugins for a template is simple:

```
$ packer init -upgrade [template/directory]
```

HCP credentials can and should be defined in environment variables. They can be defined in your Packer template code, but remember that defining sensitive credentials in your code is usually step one to accidentally publishing them to shared or public Git repositories and being compromised. These values can be found by checking the URL bar within HCP – for example, `https://portal. cloud.hashicorp.com/orgs/11eb5efd-82bd-18c1-a21f-0242ac11070b/ projects/7a4e3521-d454-4c8f-8283-2ea9f9f42b72`:

```
export HCP_CLIENT_ID=[YOUR CLIENT]
export HCP_CLIENT_SECRET=[YOUR SECRET]
```

```
export HCP_CLIENT_ORG_ID=[YOUR ORG ID]
export HCP_CLIENT_PROJECT_ID=[YOUR PROJECT ID]
```

Image data in HCP Packer has a four-tiered structure. A registry is the top of that structure, with a tree below listing builds and image versions. Tags are used to select lifecycle stages such as development and production. An HCP organization may only have one Packer registry, but as HCP is multitenant with different organizations, your user may have access to multiple registries, one in each organization. The hierarchy of registry resources may be a bit unclear at first. A registry contains buckets, which, in turn, contain a version for each build, which, in turn, contains one or more images.

> **Note**
>
> At the time of writing, HCP Packer uses the term *iteration* instead of *version*, so some of the screenshots in this chapter may be misleading.

The logical hierarchy for an HCP Packer registry should be structured like this:

1. **Registry**: The root of your image hierarchy, identified by `organization_id`.

2. **Buckets**: The container for every separate build template you use.

3. **Versions**: A list of each version for a build event.

4. **Artifacts**: The list of each artifact/image within the build version.

5. **Channels**: Images are assigned a "channel" with a release for the lifecycle.

Let's break down an example of this hierarchy for a sample HCP Packer registry added to our build. First, you will need to create a registry:

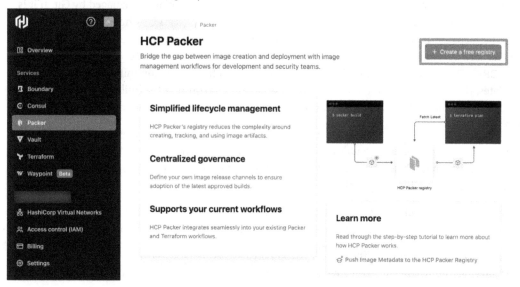

Figure 10.5 – Creating an HCP Packer registry for your organization

When a registry is created, it can be activated in your Packer template using `hcp_packer_registry`, which has a specific set of values:

```
hcp_packer_registry {
  bucket_name = "al-latest"
  description = "Latest Amazon Linux with ACME Tools"
  bucket_labels = {
    owner = "platform-team"
    os = "Amazon Linux"
    architectures = "x86_64|aarch64"
  }
  build_labels = {
    build-time = timestamp()
    build-source = basename(path.cwd)
  }
}
```

The `hcp_packer_registry` block is optional for a build block, but there can be up to one declaration. Here, we will configure what entry names are inside the registry to file this build record under.

These are bucket attributes. This may be confusing because they are not actually S3 buckets in the cloud. This field is reflected in the HCP console as `Artifact ID`, but it would be more accurately described as `Template ID`, since multiple versions may be produced by a single template as you adjust it.

The build labels will reflect in the version. It's helpful to include a map with attributes such as a build timestamp and whatever other data you find useful. There is also some implicit data that is picked up for you. Packer checks to see whether your build directory or template is managed by Git. If it is included in a repo, then Packer will automatically use its `git commit <hash>` as the version name fingerprint. If your Packer template or directory is not managed by Git, then Packer will produce an error, asking you to specify a fingerprint manually using the `HCP_PACKER_BUILD_FINGERPRINT` environment variable. The benefit of this is that if you commit no changes to your templates, the fingerprint will tell Packer not to waste time rerunning the build. If you really can't manage your templates with Git, then it is up to you to generate a hash or unique identifier for the `HCP_PACKER_BUILD_FINGERPRINT` environment variable.

Tweaking a single template as you develop it will require either setting a fingerprint or committing at each build attempt. This may sound tedious, but it will make much more sense in the next chapter when we automate builds using GitOps workflows.

Check the example template for this chapter in the Git repo: `https://github.com/PacktPublishing/HashiCorp-Packer-in-Production/tree/main/Chapter10`.

It builds two AWS AMI images. The latest Amazon Linux for 64-bit (x86_64) architectures and ARM64 (AArch64) are built with shared provisioners. For development, each test build should require a commit that provides a full audit trail.

After a fresh build is finished, there will be a new latest channel assigned. Versions within a bucket may be promoted or *tagged* via a channel such as dev or production, which can be customized. The same logic for not deploying containers with the latest tag is the reason for channels. Packer developers can adjust image templates without affecting the Terraform and DevOps teams' deployments. Channels cover the lifecycle and allow you to tag a version as **test** or **production** so that Terraform can consume only the actively selected version for that stage of the lifecycle, as shown in the following screenshot:

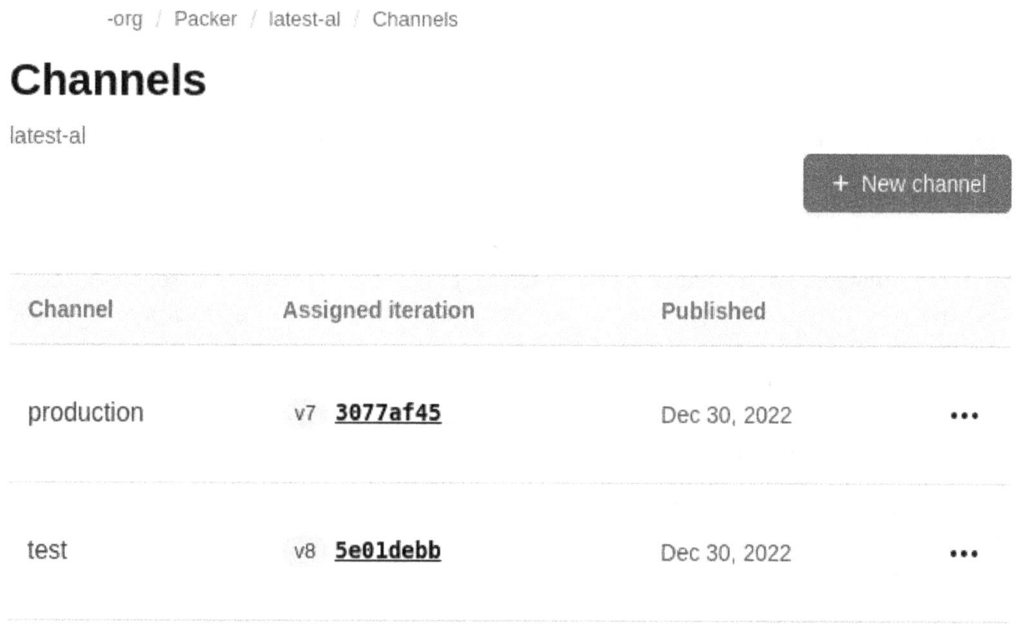

-org / Packer / latest-al / Channels

Channels

latest-al

+ New channel

Channel	Assigned iteration	Published	
production	v7 **3077af45**	Dec 30, 2022	•••
test	v8 **5e01debb**	Dec 30, 2022	•••

Figure 10.6 – Channels to tag the test and production versions (formerly iterations)

Assigning a channel to a version allows Terraform to easily look up which images are considered **production** or **test** at the time of running. Tagging or updating a channel will affect subsequent runs of Terraform and prevent Terraform from deploying old images. This is where HCP helps by providing ancestry.

Using HCP ancestry

We have discussed the concept of image hierarchy and ancestry for parent/child image relationships, but we haven't discussed how to use it. If one image serves as a base image for a hierarchy of application images, then you can retire or deprecate them all at the same time by deprecating the ancestor's channel. The way to do this is to create another template, using the hcp-packer-image data source, to look up another AMI ID to extend with further images. We will create an example where we extend the latest Amazon Linux image used in the sample template: https://github.com/PacktPublishing/HashiCorp-Packer-in-Production/tree/main/Chapter10.

We will select the development channel of al-latest but add GPU support using NVIDIA's latest drivers, tuning the environment for 3D render farm operations. The end result is to look up the current AMI ID for our region. Specifying the minimal fields for bucket_name, channel, and region would be enough, but if there are multiple images within the version, you may need to specify component_type and use the exact source.name used in the parent build. In this case, we will use the aarch64 source name to make sure we don't accidentally select the x86_64 image:

```
data "hcp-packer-image" "al-latest-aarch64" {
    bucket_name    = "al-latest"
    channel        = "test"
    cloud_provider = "aws"
    region         = "eu-west-1"
    component_type = "amazon-ebs.al-aarch64-latest"
}
```

With this information, Packer should have enough to retrieve our correct AMI. Troubleshooting errors may depend on the cloud provider, so it's a good idea to use PACKER_LOG=1 as you develop these templates. Remember that PACKER_LOG will show important trace output that may help troubleshoot issues. Data sources are not conducive to dynamic HCL or reference sources, so they will need to be created individually for each source. Here is a sample Packer source block that looks up an HCP Packer image in our registry:

```
source "amazon-ebs" "ubuntu-example" {
    source_ami = data.hcp-packer-image.ubuntu_us_east_2.id
    // More source attributes
}
```

If the version of the builder plugin supports ancestry, the resulting child/parent relationship should be reflected in the HCP UI, as shown in the following screenshot:

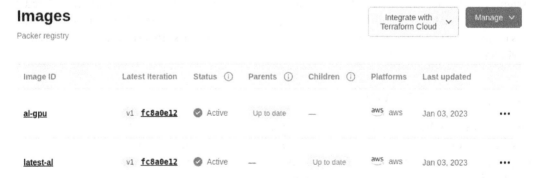

Figure 10.7 – HCP Packer ancestry from this chapter's sample code

Ancestry is not fully supported by JSON templates. The detailed feature matrix can be found on the Packer website. JSON templates may only be best applied via API and automation use cases, but feature support must be also considered. The following support matrix from the HCP documentation shows what features are supported by HCL2 and JSON templates:

Feature name	HCL2 templates	JSON templates
Basic configuration with environment variables	Full support	Full support
Custom configuration via `hcp_packer_registry`	Full support	No support
Custom image bucket description	Full support	No support
Custom image bucket labels	Full support	No support
Custom version build labels	Full support	No support
Custom image bucket description	Full support	No support
The ability to use HCP Packer data sources	Full support	No support
HCP Packer image governance	Full support	No support
HCP Packer image ancestry tracking	Full support	Limited support

Table 10.1 – HCP versus PKR.JSON HCP feature support

The way this ancestry appears in the HCP UI is shown in the following screenshot. We have a base image called `latest-al`, which is our golden image of the latest Amazon Linux release. The `al-gpu` image is Amazon Linux-built, with GPU drivers for AWS' NVIDIA offering.

al-gpu

No bucket description available.

Image details

Newest iteration	v1 **fc8a0e12** ✓ Active
Published	Jan 03, 2023, 1:31:39 PM
Platforms	aws aws
Labels	gpu:NVIDIA A10G os:Amazon Linux team:multimedia

Ancestry   Beta

Parents 1   Children 0

Image ID	Channel	Linked iteration	
latest-al	test	v1 **fc8a0e12**	Up to date

Figure 10.8 – Artifact ancestry lists

The lifecycle management of HCP allows the revocation of versions immediately or in the future. This revocation also supports ancestry. You can specify that a version expires at a future time. After this time, lookups will fail, with an intentional error saying that the image is not allowed. If the version is still assigned to a channel, then revocation is not allowed. The channel must be assigned to a new version first. An example of this is shown in the following screenshot:

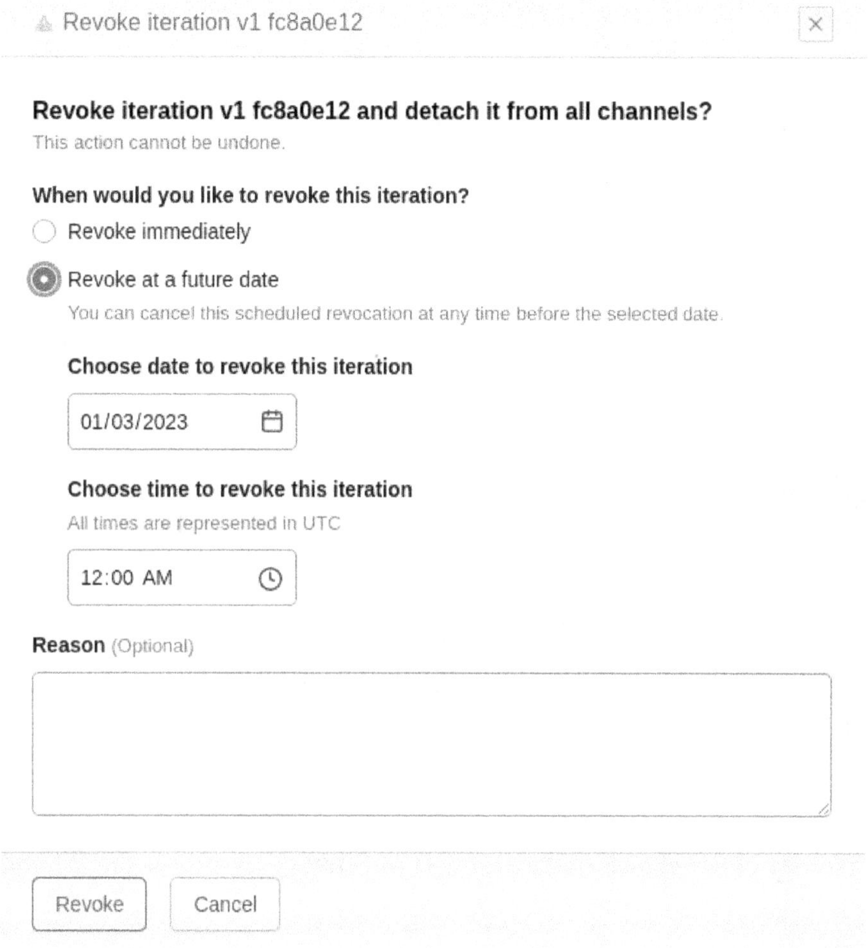

⚠ Revoke iteration v1 fc8a0e12   ☒

Revoke iteration v1 fc8a0e12 and detach it from all channels?
This action cannot be undone.

When would you like to revoke this iteration?

◯ Revoke immediately

◉ Revoke at a future date
   You can cancel this scheduled revocation at any time before the selected date.

   Choose date to revoke this iteration

   | 01/03/2023   📅 |

   Choose time to revoke this iteration
   All times are represented in UTC

   | 12:00 AM   🕐 |

Reason (Optional)

| |
| |
| |
| |

| Revoke | | Cancel |

Figure 10.9 – Revoking a version (formerly iteration) in HCP

Revoking a version prevents it from being assigned to a channel and is useful to prevent the deployment of an image that should be considered insecure or expired. Ancestry is supported if revoking a parent should revoke children at the same time. This is very helpful to make sure older, insecure images are not used. This option is found on the confirmation dialog. Note that revoking an image does not actually delete cloud storage for the image. It only affects metadata within HCP. Also, deleting the cloud storage for AMIs does not affect HCP Packer.

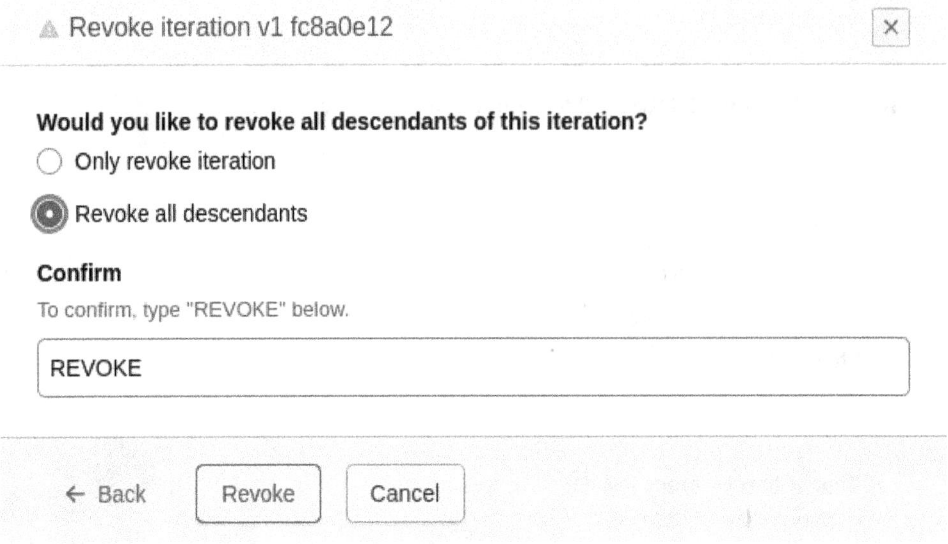

Figure 10.10– Revoking a full ancestry of images

Now, no channel may be assigned to this version, and therefore, the image cannot be looked up via a data source. Packer will not actively shut down instances using a revoked image, but it can prevent you from deploying revoked images into new instances. Now that we've got images assigned to channels and have the option to revoke an obsolete image, we should briefly cover how to actually consume this from Terraform code.

Consuming HCP Packer from Terraform

Terraform is HashiCorp's tool for infrastructure as code. Terraform is largely out of scope for this book, but we will at least cover how to use HCP Packer to deploy your images registered with HCP. The good news is that Packer's data source type actually started out as a Terraform feature. The key feature in Terraform is simply being able to look up the IDs of the AMIs built with Packer:

```
data "hcp_packer_image_version" "gpu-test" {
  bucket_name = "al-gpu"
  channel     = "test"
}
```

If this looks familiar, it's because the Packer data source and the Terraform data source use the same syntax and the same API behind the scenes. Now, as you update your release channels in HCP Packer, those changes will automatically be picked up by subsequent Terraform runs:

```
data "hcp_packer_image" "amiid" {
  bucket_name    = "al-gpu"
  cloud_provider = "aws"
```

```
    region          = "eu-west-1"
}
```

Just like in Packer, we must specify a channel name or version ID from the data source. If looking up multiple images in a version, it's wise to use `version_id` instead of a channel. Since the sample code contains images for x86_64 and aarch64, we'll use a version ID to look up both AMI IDs. If you do wish to use just the channel and not specify a version ID, then it looks like this:

```
    channel = "test"
```

The `channel` option is fine, but the provider will need to look up a version ID for each image. In this sample, we use `version_id` directly to save some time. Every bucket in the chain must have a matching channel established for this to work:

```
    version_id = data.hcp_packer_image_version.gpu-test.id
```

The sample code for this chapter simply outputs the AMI IDs for both images in this version. That includes `amiid-x86_64` and `amiid-aarch64`. A successful `terraform apply` command inside the directory will generate some output and save a simple JSON Terraform state file into the current working directory. A `terraform apply` command will save the latest output values to the Terraform state without changing any real infrastructure or deploying an instance:

```
$ terraform apply --auto-approve
data.hcp_packer_image_version.gpu-test: Reading...
data.hcp_packer_image_version.gpu-test: Read complete after 0s
[id=01GNVX96F1J9S0QE10EKFJXW12]
data.hcp_packer_image.ami-x86_64-gpu: Reading...
data.hcp_packer_image.ami-arm-gpu: Reading...
data.hcp_packer_image.ami-arm-gpu: Read complete after 1s
data.hcp_packer_image.ami-x86_64-gpu: Read complete after
Changes to Outputs:
  + amiid_aarch64 = "ami-0049287656a99b108"
  + amiid_x86_64  = "ami-0cad1567aca722891"
Apply complete! Resources: 0 added, 0 changed, 0 destroyed.
```

You can even use `jq` to inspect the state file:

```
$ jq -r .outputs.amiid_aarch64.value terraform.tfstate
ami-0049287656a99b108
```

There are also other useful outputs from the data source. You can fetch details about the HCP organization or the build. You can also get the labels and the `revoke_at` time for the image if the `revoke_at` time is specified for the image. It would be wise to make a note of the `revoke_at` time inside your image somehow. Whether this is in the MOTD, an asset manager, or a lifecycle management system is up to the developer. If you're wondering how Packer communicates with HCP or whether the API can be used directly, we will cover the basics of the HCP Packer API next.

Exploring the HCP REST API

The Terraform provider for HCP can help automate HCP configuration, but if there is a need to access the API directly, the good news is that Packer is one of the simplest and best-documented of HCP's API endpoints. To begin interacting with any HCP API, you need to first get a bearer token using your service principal credentials. You can use the same environment variables as the Terraform examples for a simple `curl` statement:

```bash
#!/bin/bash
curl --silent --request POST \
  --header "Content-Type: application/json" \
  https://auth.hashicorp.com/oauth/token --data @- << EOF
{
  "audience": "https://api.hashicorp.cloud",
  "grant_type": "client_credentials",
  "client_id": "$HCP_CLIENT_ID",
  "client_secret": "$HCP_CLIENT_SECRET"
}
EOF
```

On completion, this will return a JSON document containing details of a short-lived token:

```
{"access_token": "[YOURTOKEN]", "expires_in": 3600, "token_type":
"Bearer"}
```

This token can then be used to access the HCP endpoints using the privileges from the service principal originally authenticated. The HCP API endpoints vary by product and release version, so be sure to check the HCP documents for the project's current information.

The documentation follows the OpenAPI V3 formatting but doesn't always describe using examples. There is one more environment variable that the Terraform provider uses, which can also be used for your API access. That is `HCP_API_HOST`, which contains the HCP service to access. By default, the API endpoint will be `https://api.cloud.hashicorp.com`. The organization ID and project ID are required for most calls and can be fetched from the HCP UI or by using the core HCP API. For example, to fetch a list of your organization's buckets, use this:

```
curl -H "Authorization: Bearer $HCP_ACCESS_TOKEN"
$HCP_API_HOST/packer/2021-04-30/organizations/<location.organization_
id>/projects/<location.project_id>/images
```

The API may be useful to report dashboards or automation with external systems that don't provide native CLI functionality. There are also community integrations for most HashiCorp APIs and even community FUSE filesystem drivers. HashiCorp services and HCP services can be browsed like a filesystem locally, although Packer is fairly new to HCP and not all features can be implemented by community projects.

For full documentation of the HCP Packer API, please check the HashiCorp Developer portal: `https://developer.hashicorp.com/hcp/api-docs/packer`.

Summary

In this chapter, we discovered how to use HCP Packer to manage the lifecycle for us. The mechanisms to produce and consume HCP Packer data are similar for Packer and Terraform in delivering a consistent experience. The HCP Packer API is evolving with new features, such as ancestry and family revocation, so be sure to check the documentation pages for the latest features. Also, make sure that the builder plugins you use support the HCP Packer features required and that the latest versions of these plugins are used.

In the next chapter, we will take everything we have done manually up to this point and apply automation. Automation will make Packer development much easier, as you don't need to manually run build steps, and you will have checks to prevent you from making common mistakes as part of your process.

11
Automating Packer Builds

In the previous section, we covered the lifecycle management of images and how to make sure you retain historical data about images for archiving, forensics, or auditing. Now, we're going to put together everything we've learned and apply automation using a full GitOps workflow. Automation really should be applied early on in your Packer journey, but it's important that we've covered all of the basics of manual builds first to understand how to properly automate them. Some of the tools we created earlier, such as scripts to build all images in a directory, are perfectly set up for basic **Continuous Integration** (**CI**) pipelines via tools such as Jenkins, GitLab CI, and GitHub Actions. There are many tools to cover, but we will first cover some fundamentals of automation that must be considered in any automation solution. Then, we will cover some helpful examples using widely available open source tools that are commonly used in enterprise environments.

We will cover the following topics in this chapter:

- Identifying common automation requirements
- Exploring basic GitHub Actions support
- Exploring GitLab CI pipeline support
- Using HashiCorp Vault integration for pipelines

By the end of this chapter, we should be capable of fully automating everything we've covered in Packer so far, which will greatly help speed up any development process. The dual acronym **CI/CD** refers to **continuous integration and continuous deployment**, which are often both covered by a single solution. Recent trends further combine these features with the **Version Control System** (**VCS**). GitHub and GitLab store code and provide version control workflows but now also provide basic CI functionality to perform build and test actions. These features are very helpful for Packer development.

Note our focus on CI covers building and integrating combined environments with Packer but does not cover CD and actually deploying a VM or container workload to production. Deploying a production image from Packer is usually performed via a Terraform workflow or other CD tools, such as HashiCorp's Waypoint project. For now, all we need to automate is the workflow where changes to Packer code result in new images as quickly as possible. We will cover this using GitHub and GitLab examples. Both of them are fairly similar in that they use YAML files to invoke build jobs based on commit events. Both of them also support free SaaS tiers that can be used to follow along with the examples in this chapter. These build processes will be actioned by a listening service called a runner. First, we need to gather some requirements to decide what type of runner is necessary.

Technical requirements

In this chapter, we explore automation with widely available SaaS platforms (`GitHub.com` and `GitLab.com`). You should only need access to the Internet and a cloud account to try the examples for this chapter. Knowledge of Git is recommended or you may just edit your Packer templates via the UI these products present in a web browser.

Identifying common automation requirements

Establishing a runner depends on what types of images you'll be building and where. Runners will connect to your GitHub or GitLab instance to monitor a repository for event hooks. The advantage of this is that no ingress ports must be open on your network or firewall. A runner service will simply establish secure API access to your GitHub or GitLab instance and listen for events such as a merge or commit on a certain branch. This is the most basic form of CI, which is actually sufficient for most Packer tasks.

What type of runner you need depends on what types of images you will build. Building cloud images often relies only on API calls and can be done from just about any environment. A runner service may be a container, a VM, a physical machine, or a managed service provided by GitHub or GitLab.

If building remotely in a cloud provider, any runner should suffice. Since the builder is only making API calls to external resources, not much compute or RAM is required. Often, the basic managed runner environment should suffice. We will start with the cloud use case for the first example in this chapter because there is no runner configuration required.

Containers are easy to build with a VM, as virtualization hardware support is not required for containers. Using a container push postprocessor to publish images to a registry will minimize bandwidth and storage requirements, as image layers are deduplicated based on checksum. If 50 different container images are built, only the differences will be uploaded.

A bare metal runner may be used if you're building or testing local VM resources or mixed architecture images. A great example of this is to cross-build and test your application for mixed architectures locally without any cloud costs. This development use case can be applied across application development pipelines as well as Packer or cloud DevOps pipelines. A single change to either application code or Packer templates can trigger a cross-build and test of your application using QEMU VMs on x86_64, ARM, AArch64, Power, S390, RISC-V, and more all at once with a simple development machine.

For the example code in this chapter, we will offer examples in two categories for both GitHub and GitLab. Each VCS option will offer cloud examples for AMIs and local examples using the versatile QEMU builder. These examples are modified versions of previous content from earlier chapters. Use whichever example suits your options just so long as you are able to test the automation applicable to you. For mixed architectures, it is wise to add a self-hosted physical runner tagged *QEMU* or whatever builders the host supports. Then, you can use specific pipelines for appropriate builds. You can have a local QEMU runner build your ARM or Raspberry Pi images for IoT or set-top boxes while having cloud runners build your cloud infrastructure images.

Multiple runners may be configured on a single repository or project if you need hybrid cloud support. One self-hosted runner may handle local VMware builds while a cloud or managed runner may handle cloud resources. This helps protect local VMware endpoints that don't need to be exposed to your VCS solution as they only need access from the runner actually performing builds.

Be sure to consider parallelism in your automation pipelines. By default, most of these pipelines will trigger immediately for any commit or merge to your selected branch. This may result in overlapping pipelines and possible resource conflicts. Hardcoding a VM name may cause issues if two or more instances of Packer are run. If your CI platform allows the option of serial pipelines or interrupting any running pipelines before stating the latest, that is a convenient option. There is no need to finish a build pipeline if it's already not using your latest code because there was another commit on the current branch. If your CI platform does not offer this option and you are running a dedicated builder, then you may want to take advantage of Packer's SIGHUP handler. Running this at the beginning of every pipeline can tell any running instance of Packer to gracefully stop, freeing up all resources and flagging the pipeline as stopped by the user so that the latest pipeline can run. It's important to handle a failure result of the `pkill` command so that the pipeline keeps running if there are no running `packer` processes:

```
pkill -u $USER packer || echo "No packer process stopped"
```

Depending on your lifecycle strategy, there may be a desire to automatically redeploy dev instances every time an image is rebuilt. Deployment can be added as part of this automation. Pipelines accommodate a `terraform deploy` command or a call to a script for updating environments with Terraform or HashiCorp's Wayland deployment pipelines. For now, we will leave placeholders for deployment as a comment in the sample code.

In this chapter, we cover examples with GitLab and GitHub, which combine version control and CI. Not all solutions do this. It's convenient to have both solutions combined. If forensics or testing teams need to reproduce an issue from last year and there is no image available, it's super simple to just pull up the pipeline from last year, which can be re-run at any time, and reproduce a historic image from code. This is likely an option with separate version control and CI solutions too. If the two systems aren't maintained together or one system introduces a regression during updates, then problems can occur. Compatibility with historic Packer jobs can be problematic to salvage.

Cost may be a consideration also. GitHub's and GitLab's free versions support automation, but managed builders may not be free. Self-hosted runners are a good way to save costs. Many organizations will already have full-featured and supported accounts for a combination of features or capacity, but if you are learning on your own self-hosted environment, local runners save the platform's allocating capacity to you. Sometimes, basic accounts will only allow a certain quantity of managed runner operations. If you run into this trying to reproduce the examples for this chapter, make sure to try self-hosted runners.

Exploring basic GitHub Actions support

To work with GitHub Actions, you need a GitHub organization and admin access to your GitHub repository. You may choose to build your own repository from scratch or you can fork the book's sample repository if you haven't already. Now, you will have access to the Actions menu for your repo. The examples and screenshots in this section are taken from setting this up on the book repo.

GitHub supports multiple runners on a single repository. GitHub makes it simple enough to add a self-hosted runner, so let's try that first. Here, GitHub will guide us to download the appropriate binary for our selected platform and architecture. macOS, Linux, and Windows are supported:

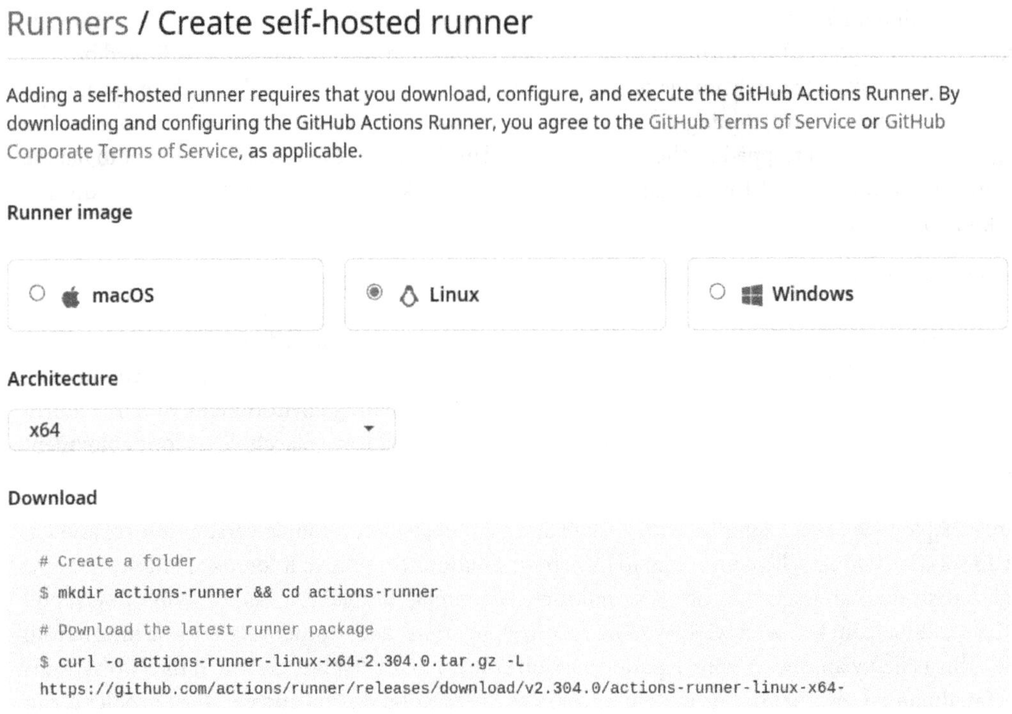

Runners / Create self-hosted runner

Adding a self-hosted runner requires that you download, configure, and execute the GitHub Actions Runner. By downloading and configuring the GitHub Actions Runner, you agree to the GitHub Terms of Service or GitHub Corporate Terms of Service, as applicable.

Runner image

○ macOS	◉ △ Linux	○ Windows

Architecture

x64 ▾

Download

```
# Create a folder
$ mkdir actions-runner && cd actions-runner
# Download the latest runner package
$ curl -o actions-runner-linux-x64-2.304.0.tar.gz -L
https://github.com/actions/runner/releases/download/v2.304.0/actions-runner-linux-x64-
```

Figure 11.1 – GitHub guide for creating a self-hosted runner

GitHub currently distributes standard archives for Linux platforms without distribution packaging, so be careful which download you get. Make sure to use the latest version at all times because these runners will configure and save some important sensitive API tokens for accessing GitHub. Ensure you install and configure this runner with security in mind. This means the following:-

- Create a dedicated user or container with minimum permissions.

- Protect config files and tokens from other users.

- Write artifacts and logs to secure storage.

- Never copy or back up the entire directory to a vulnerable location.

When downloading the GitHub runner binary, GitHub will automatically present a configuration command with a temporary setup token. In the repo, go to the **Settings** | **Actions** | **Runners** section to list or add a self-hosted runner.

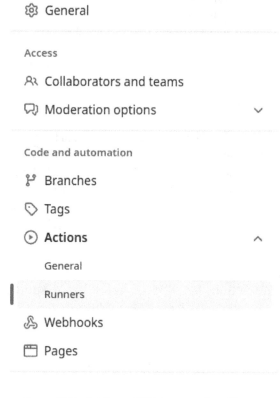

Figure 11.2 – Adding a GitHub runner in settings

For example, when setting up the action on the repo for this book, the command looked like this. The organization and repository are automatically populated and a temporary token appears:

```
./config.sh --url https://github.com/ORG/repo \
    --token [TEMPTOKEN]
```

Running this creates some very important hidden files with your credentials. Be aware that copying this entire directory will include these credentials. Permissions are set to minimize access only to the user who ran this setup:

```
$ ls -lhta .credentials*
-rw-r--r--. 1 user group  266 .credentials
-rw-------. 1 user group 1.7K .credentials_rsaparams
```

Note that the .credentials file contains non-sensitive identifiers but .credentials_rsaparams contains sensitive RSA keys for decrypting your identity and authenticating it to GitHub. These keys are what the runner uses to connect to GitHub. Once this completes successfully, you may run the actual runner service or configure it as a systemd unit so it will always run in the background:

```
$ ./run-helper.sh
√ Connected to GitHub
Current runner version: '2.300.2'
2023-01-15 15:06:12Z: Listening for Jobs
```

Now, the **Runners** section of your repository config in GitHub should show the runner's status. If you stop the service, it will go idle inside GitHub after a brief timeout. We could include custom labels such as physical or test, but notice the default labels include self-hosted, Linux, and x64. These labels can be used to select a specific runner through our pipelines.

Figure 11.3 – Runner idle on a local machine

Now that we have a runner enabled, we can declare a pipeline. Head back to the **Actions** menu of your repo. Remember you need administrative rights to your repo for accessing the **Actions** menu. Now we will add a new workflow. GitHub and GitLab both make this simple with helpful templates based on typical use cases. At print time, GitHub has no template for Packer but we can make our own. GitLab has a simple template for Packer. A generic template is fine because we will simply make use of the Packer build scripts we wrote earlier. Have a look at the **Actions** settings tab in the following screenshot:

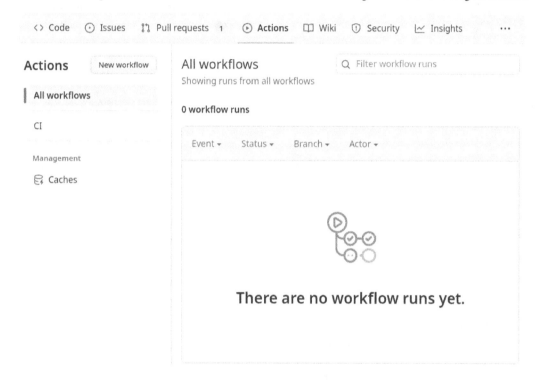

Figure 11.4 – Ready to add workflows

Now, let's add our workflow. This will consist of super simple YAML to get going. We will need to specify what repo action triggers a workflow and what type of runner we need to schedule it on. These YAML files become part of our repository, which allows them to be version controlled in addition to our source code. For this chapter, we will create two separate pipelines. One of them uses self-hosted runners and one uses a Ubuntu container managed by GitHub. These could be combined into a single workflow but we will leave them separate for demonstration. Full examples can be found inside the book's GitHub repository: `https://github.com/PacktPublishing/HashiCorp-Packer-in-Production/tree/main/.github/workflows`.

There are two main differences to consider between the sample pipelines:

- What paths within the repo we trigger workflows for

- What type of runner can handle each workflow

First, identify the paths within your repo you wish to trigger on. In this case, the only reason to trigger a workflow is a change to the `Chapter11` folder. Since we have separate directories for self-hosted and cloud-hosted runners, we will have one of them trigger on `Chapter11/self-hosted/*` and one on `Chapter11/cloud-hosted/*`. A partial example of this YAML is as follows:

```
on:
  push:
    branches: [ "main" ]
    paths:
      - Chapter11/self-hosted/*
```

Next, check the `jobs.build.runs-on` value. To trigger this pipeline only on a runner with certain tags, we can select self-hosted runners or specific OS/container combinations. In this case, we need a list of tags, which implies all tags must be present to qualify: `[self-hosted, linux]`. Now, this will run on any self-hosted Linux runner. If you find workflows wait indefinitely for a runner to pick them up, make sure your runners and workflow tags match:

```
jobs:
  build:
    runs-on: [self-hosted, linux]
```

This workflow will only trigger on self-hosted Linux runners. Note that this does not specify physical or virtual environments, so that would be up to us to tag in our runner. Now, a packer build will be attempted on each change of the folders within the `Chapter11` directory on a push or merge. If the local runner is offline, the job will wait for up to 36 hours (by default) before registering a timeout.

It may take several attempts to perfect your pipelines, but in the end, having a stable automation pipeline is worth it. Now, any change to these directories on GitHub will trigger build pipelines in a way that tracks the identity of who committed the code and what the build result is. A successful build can commit the latest images to the cloud or to container registries where they can be consumed or tested. Make sure all security is considered with the logs. Don't let a build use any long-lived credentials that might be written to logs during a build. Ideally, HashiCorp Vault is used to provide short-lived, self-destructing credentials that won't be available outside of any build environment. If not, make sure that the logs are hidden for a period of time long enough that nobody can view sensitive credentials. Public repositories in a SaaS VCS such as GitHub.com or GitLab.com are common places for hackers or bots to troll for secrets. Always use a private repository for automation or use very tight security policies of minimum privilege on your repo.

By now, we've established a basic setup for GitHub Actions to perform automated Packer builds whenever you make code changes. Be sure to check the GitHub documentation for the latest information if processes change. Feel free to experiment with stages of the pipeline and Packer templates. Also, experiment with different types of runners. As long as a runner environment has Packer installed, it should be possible to test most builds. For advanced local emulation and QEMU builds, a physical runner is best.

Exploring GitLab CI pipeline support

Now that we've seen how GitHub Actions handles basic automation, let's repeat the same experiment for GitLab CI. GitLab offers a free tier of SaaS and also a free edition of the private GitLab Community Edition. GitLab offers the paid supported Enterprise platform edition, which is recommended for production workloads. Enterprise users have access to additional features than what Community Edition contains. Both the SaaS GitLab.com and private GitLab options can serve to follow the examples in this chapter. With an organization and admin access to a repository in GitLab, you will find GitLab runners are almost identical to GitHub runners. The user has the same option of managed or self-hosted runners. First, we need at least one runner to actually perform Packer builds. Similarly to GitHub, download and install the runner type needed. This can be a local self-hosted runner or a cloud or managed runner just like GitHub. GitLab has a bit more detail provided in the YAML manifests and there is an example provided in the repo directory of this chapter.

Self-hosted GitLab runners offer standard repositories and Linux packaging, which offers a standard runner binary. A FIPS version of the runner that supports FIPS compliance for financial environments is also available. The FIPS runner also supports the FIPS kernel mode, which is nice. The general runner should be fine for most users who don't have FIPS 140-2 requirements or specific encryption ciphers. Have a look at the following runner configuration screenshot:

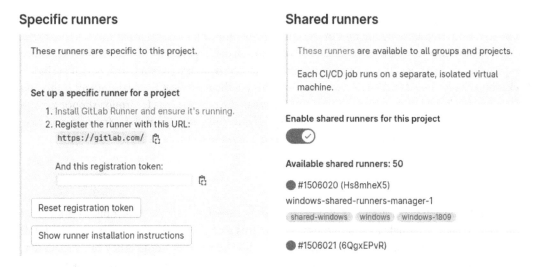

Figure 11.5 – GitLab runner deployment

Install the GitLab runner package for your platform and then register it with GitLab's helper command. The GitLab instructions will provide you with a token for your organization. Make sure the token does not remain in your shell history. Adjust the runner if necessary, including the user used to run the systemd service provided. It seems to run as root by default, which should be changed right away to a dedicated and secure user. You may create one if needed. If you are running your own GitLab instance, make sure to adjust the URL appropriately:

```
gitlab-runner register --url https://gitlab.com/ \
    --registration-token [YOURGITLABRUNNERTOKEN]
```

GitLab runners allow specifying an executor. Each pipeline can be run in a new Docker container or Kubernetes Pod or a number of executors. Large-scale environments with hundreds of parallel runs may benefit from a Docker or Kubernetes executor but most times, the simple *shell* executor is perfect. This runs the build commands locally in the runner environment similar to GitHub's runners.

GitLab CI pipelines are defined with simple YAML. In fact, GitLab has an existing template for Packer CI pipelines. This helps generate the GitLab version of the GitHub action we built earlier. The template itself is hosted in a public GitLab repo and is open to public merge requests if the community wants to edit or improve it.

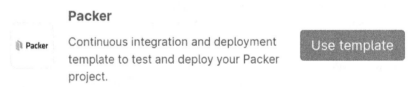

Figure 11.6 – GitLab CI's Packer template

This template includes pre-checks and pipeline steps to verify a working Packer binary and environment before iterating over all template files in the current directory files and building them serially. It is possible to customize the template and run the same command in parallel or to find files using a more complex search. First, we find the pre-check in the `before_script` root step, which verifies a Packer binary is installed in $PATH. Not only does this fail the pipeline if the environment is missing a Packer install but it will also log the version of Packer to the pipeline logs for troubleshooting. Note for best security practices, it may be preferred to specify the full path, `/usr/bin/packer`, in case someone is able to inject a different binary as `packer` into the runner's $PATH:

```
before_script:
  - packer --version
```

Simple validation will make sure your Packer code can be interpreted by the Packer binary. It's a good idea to add a packer validate as a pre-commit hook. This isn't part of GitHub Actions or a GitHub CI pipeline but part of a Git workflow policy. GitLab calls these **server hooks**. They can declare that any change to a `*.pkr.hcl` file should apply a `packer validate` command to make sure the syntax is valid before any merge is performed. This will prevent wasted automation pipelines from obvious failure.

If you want to test your code against multiple releases of Packer and you're using Bash, there is a handy trick to loop through all versions of Packer in your environment for each of these stages:

```
for p in $(compgen -c | grep '^packer')
do
  # Substitute the appropriate packer command here:
  eval "$p version"
done
```

Note the security implications of using `eval` and ensure every version of `Packer` in your path has been installed from packaging with verified signatures which guarantee the proper developer build is the one being used. If you copy and append Packer versions with the release version, such as `packer_1.8.4`, make sure that the binary is verified before using automation. It's wise to go a step further and supply the full path to each Packer binary. If you don't want to use the Packer template from GitLab, you can manually create a `gitlab-ci.yaml` file to your repository:

```
build:
  stage: deploy
  environment: production
  script:
    - find . -maxdepth 1 -name '*.pkr.hcl' -print0 | xargs -t0n1
packer build
```

Define which stages are included in this YAML. In this case, we will implement the validate and build actions. If you want to have extra events to test or deploy, you can actually add scripts to continually test your applications or even deploy the updated images triggering a `terraform apply` command or a Terraform Cloud run process. Unfortunately, Packer has no record of Terraform workspaces that use each image output. It would be up to you as a developer to identify the appropriate Terraform workspaces to apply changes made to your Packer build outputs.

The previous basic simple example will build all packer template files at the root level of your repository serially one by one. If any error occurs, the entire pipeline will stop and reflect a failure, which helps with troubleshooting. At a second glance, this code presents a few problems. What if you include non-Packer directories in your repository? Or what if you want directories nested deeper than one level? In this case, it may help to use a special suffix in your directory naming to indicate which directories actually contain Packer code. Changing the `find` command makes this easy, as follows:

```
script:
  - find . -type d -name '*.pker.d' -print0 | xargs packer \ build
```

Now, all directories in your repo ending with `.pker.d` will be built no matter how deep they are in your repository structure. They will still be built one at a time, unless `xargs` is changed to parallel mode (-P #) or GNU parallel is used instead. The find command by default traverses directories and their subdirectories first before moving onto directories at the base level. The find command can also do a depth-first traversal, meaning that deeply nested directories may be built first before moving on to the next directory at each level. Remember the discussion regarding structured directories for image dependencies. A breadth-first traversal where each layer of directories is built can be scripted with a recursive function:

```
buildall()
{
 for d in *.pker.d
 do
   packer build $d
 done
 pushd "$d"
   buildall
 popd
}
```

In this case, the build command is more than one line, so it is best to place it inside a script. The build script can be specified rather than inlined. Also, another challenge comes to mind. How long should a pipeline run before it is considered stuck and canceled? Let's specify a max time for a pipeline to build. This will prevent hanging pipelines or accidental infinite loops in a Packer build due to a never-ending pipeline:

```
build:
   script: build.sh
   timeout: 1 hours 30 minutes
```

This will let a build run for up to 90 minutes before killing all processes and flagging the pipeline as failed by timeout. It may be necessary to increase this timeout for extremely complex builds, though 90 minutes should be long enough for most builds by far. You can see GitLab provides a simple set of tools to help automate your Packer builds and do even more. GitLab can also apply Terraform or other commands that are outside the scope of this book.

Using HashiCorp Vault integration for automation

When you use Packer in pipelines it may be useful to add HashiCorp Vault for short-lived credentials. When building cloud images, shouldn't each build pipeline use its own cloud credentials and purge them afterward? What if Packer or part of the pipeline fails and doesn't get a chance to revoke any credentials used? HashiCorp Vault is there to rotate and revoke your unused credentials.

If the runner service you select is a VM within the cloud, then Packer may automatically use the service account presented to the VM itself. In this case, there may be no need to use Vault credentials. AWS, Azure, and GCP all support this option. If you are using a local runner or building for multicloud, Vault is the best way to provide secure credentials for Packer pipelines.

This section builds on the Vault discussions from *Chapter 3, Configuring Builders and Sources*. It assumes a basic knowledge of HashiCorp Vault but focuses solely on one authentication method and the secrets engine for dynamic cloud credentials. The security given by Vault can prevent someone from obtaining your sensitive cloud credentials from a log or error message. Here, we will set up a basic use case that meets the following requirements:

- Every Packer run or pipeline will use its own credentials for cloud connections
- The credentials will have a **time to live** (**TTL**) of five minutes
- Environment variables will be used according to builder specifications
- Nobody should have access to build logs until after all credentials are revoked

There are a few different authentication methods that can be considered for automation. Common methods for automatic authentication into Vault are cloud service accounts, Kubernetes service accounts, JWT/OIDC, and AppRole. The common authentication methods and secrets engines for Vault can be found in the following diagram showing authentication methods on the left, authorization via policies in the middle, and secret engines on the right:

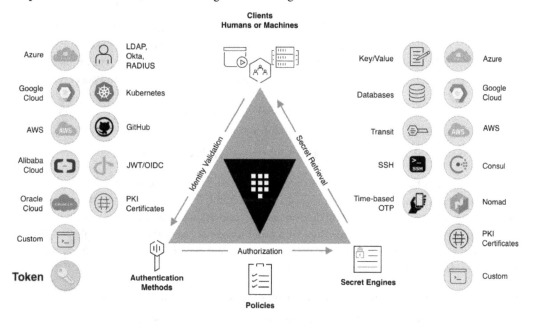

Figure 11.7 – Common authentication methods and secret engines in HashiCorp Vault (Source: HashiCorp)

Each pipeline can automatically access Vault to retrieve secrets and credentials using the built-in Vault secret integration of Packer. Some CI options have documented native support for Vault authentication. GitLab Premium and GitHub support Vault's JWT/OIDC authentication method and automatically inject a **JSON Web Token** (**JWT**) environment variable into each pipeline. GitHub actually has a dedicated Vault authentication method to simplify this further.

If your runner environment has the Vault binary installed, it's simple to use Vault commands directly for authenticating and accessing secrets. If not, then Vault's REST API can be used for everything described here. No matter how you access Vault, it is required to have at least the following two environment variables: `VAULT_ADDR` and `VAULT_TOKEN`. The `VAULT_ADDR` environment variable is usually not considered a sensitive value and can be defined globally. `VAULT_TOKEN` will need to be obtained by authenticating to Vault. Next, we'll highlight how to do this for GitLab CI and GitHub Actions.

GitLab CI

A generic pipeline with no native Vault support can use the AppRole authentication mechanism. This is a generic way to embed a token for Vault authentication into an external secret manager, such as a GitLab CI secret or a GitHub Actions secret. We will use documentation from GitLab and GitHub, but please make sure to check the latest documentation from these vendors using these links:

- `https://docs.gitlab.com/ee/ci/examples/authenticating-with-hashicorp-vault/`

- `https://github.com/hashicorp/vault-action`

The GitLab and GitHub documentation links are not Packer specific. Pipelines can use Vault for any automation, but we will be focusing on the Packer use case. A lot of use cases for Vault use static secrets. These are basic CRUD secrets, such as a database record that can be read or written. The problem with static credentials is liability. Imagine a scenario where a team of two is working with Packer on AWS:

1. Person 1 creates AWS credentials and puts them into a box for person 2.

2. Person 2 creates a Packer pipeline in a GitHub action.

3. Sudden abuse or attacks start occurring with Packer's AWS credentials.

Depending on the permissions allocated to your cloud credentials, someone may hold your data for ransom or begin mining cryptocurrency with your account. It may actually even be person 1 or person 2 who is doing this. Forensics will open an inquiry and ask who has seen the credentials. Both person 1 and person 2 have seen them. They both point fingers at each other. As multiple people on the team have seen these credentials, they are both suspect and it's difficult to prove who is at fault. Now compare this to Vault's dynamic credentials:

1. Person 1 configures Vault with AWS credentials to issue other AWS credentials.

2. Person 1 creates a role for Packer that generates five-minute credentials.

3. Person 1 configures Vault OIDC authentication for the Packer GitHub action.

4. Person 2 develops Packer templates via GitHub and never sees any credentials.

In this case, neither person 1 nor person 2 has seen Packer's credentials. Every pipeline in GitHub fetches a new random credential from Vault. Each credential is revoked in five minutes or renewed every five minutes until the build is done. When no Packer pipeline is run, there are no active AWS credentials for Packer.

Vault behaves like a DHCP server for secrets. Instead of getting an available IP address good for a lease time with the option to renew, Vault will issue randomized credentials good for a short lease time with the option to renew. This guide assumes you have a Vault instance available. If you don't have a Vault instance, you can run a dev Vault instance in memory or get a hosted dev or production instance from HashiCorp Cloud Platform. There is a cost associated with HCP Vault since it creates VMs in the cloud: `https://www.hashicorp.com/products/vault/pricing`.

In *Chapter 3, Configuring Builders and Sources,* we covered how some builders use the dynamic secrets within Vault. Standard environment variables and paths may be used for cloud authentication. The focus here is the JWT authentication method. We will need to perform four tasks to set up Vault for JWT authentication for GitLab:

1. Enable JWT authentication.

2. Configure the JWT token reviewer.

3. Configure a role for Packer.

4. Configure a Vault policy for the role with read access to cloud credentials.

This is pretty straightforward. This example is adapted from GitLab's documentation. Adjust the hostnames for your own GitLab instance:

```
$ vault auth enable jwt
$ vault write auth/jwt/config \
    jwks_url="https://gitlab.example.com/-/jwks" \
    bound_issuer="gitlab.example.com"
$ vault write auth/jwt/role/packer - <<EOF
{"role_type": "jwt",
 "policies": ["packer"],
 "user_claim": "packer",
 "bound_claims":
 { "project_id": "1",
   "ref": "main",
   "ref_type": "branch"
 }
}
EOF
```

This will only authenticate a JWT with bound claims for the main branch. It may be a good idea to have separate roles for development and production. Now, Vault needs a role to say what secret paths Packer has access to:

```
$ vault policy write myproject-staging - <<EOF
path "gcp/token/packer" { capabilities = [ "read" ] }
EOF
```

Once Vault is configured and the GitLab runner is able to authenticate to Vault and set the VAULT_TOKEN environment variable, Packer's GCP builder has the built-in feature vault_gcp_oauth_engine option, which can be added to your Packer template (HCL2 example):

```
vault_gcp_oauth_engine = "gcp/token/packer"
```

If properly configured, build pipelines will now automatically fetch a temporary cloud token for use during each Packer build. Remember, this feature requires the GitLab Premium tier. If you don't have access to GitLab Premium, you may still keep your GitLab runner in the cloud to authenticate to Vault via a cloud service account. There are many options for authentication in Vault, so if you want to learn more about Vault, be sure to check the latest Vault documentation.

GitHub Actions

Now that we've covered GitLab CI, what about GitHub Actions? GitHub has a dedicated Vault authentication method for this very use case. It is a variant of the JWT authentication method that is specifically used for GitHub Actions pipelines. The runner will automatically get a token via the GITHUB_TOKEN environment variable, which can be used to log in to a Vault instance if the instance is configured to authenticate it. The GITHUB_TOKEN environment variable can be used to authenticate against Vault to set Vault's VAULT_TOKEN environment variable. The steps to configure Vault are similar to GitLab. First, enable authentication, then configure the authentication and roles. In this case, the GitHub teams can be mapped to a policy instead of creating a role from scratch:

```
$ vault auth enable github
$ vault write auth/github/config organization=YOURORG
$ vault write auth/github/map/teams/packer value=packer
```

Now, it's simple enough to set the VAULT_TOKEN environment variable inside your build pipeline. Just authenticate via the CLI:

```
$ vault login -method=github token="$GITHUB_TOKEN"
$ export VAULT_TOKEN=$(cat ~/.vault_token)
```

Or use the API and set VAULT_TOKEN using the output of this command:

```
$ curl -X POST --data "{\"token\": \"$GITHUB_TOKEN\"}" \
    http://yourvault:8200/v1/auth/github/login \
    | jq -r .auth.client_token
```

Once VAULT_ADDR and VAULT_TOKEN are set, Packer can use standard Vault integrations. Vault Enterprise users may consider additional values, such as VAULT_NAMESPACE, and you may require a proxy to access Vault. All of these features can be found in the Vault documentation.

Summary

Don't be discouraged if it takes a while to perfect your workflows. A lot of trial and error is part of the development process with Packer templates and CI automation. The important thing is to get comfortable with an automated workflow that helps make minor changes to Packer templates without manually running builds along the way.

The sample code for this chapter shows examples of GitHub Actions and GitLab CI pipelines. They can be browsed and customized, but to try these yourself, you will need to fork the repo or create your own in GitHub or GitLab and add your own runner where appropriate. Runners use a pull mechanism that can enable a public or a private VCS platform such as GitHub or GitLab to trigger automatic builds even behind firewalls or private networks.

Automation provides instant collaboration and audit trails for a team to edit Packer templates using a GitOps workflow and view the full history of logs for any build in the past. As the logs from these automated builds may contain sensitive information, it's best practice to use Vault for short-lived credentials in your builds where possible and also to limit who can view build logs with a minimum privilege policy. Here, we added Vault authentication methods that are available for use in automation pipelines. Packer shouldn't need to find a way to authenticate to Vault if the automation can already do this for Packer.

In the next chapter, we will change gears a bit and consider how to extend Packer via plugins. You can use the automation skills from this chapter to continuously test your Packer plugins.

12

Developing Packer Plugins

In the previous chapter, we covered basic automation for Packer builds. Up until now, most of Packer's existing builder use cases should be familiar, but what about the functionalities Packer lacks? Since Packer is open source, you could fork and edit the code base, but it's much easier to write a plugin. HashiCorp's projects use Go and **Remote Procedure Call (RPC)** for plugin development, which allows plugins to be developed as standalone Go binaries or to be built into the project itself. Plugins come in four flavors: builders, provisioners, post-processors, and data sources. Each function is defined by an interface that must be implemented. Packer makes this easy to do by providing templates. We'll explore the templates by implementing two plugins in this chapter. First, we'll implement a basic data source and then a more complex builder. Before any of that, we will cover the basics of Go in the context of Packer. This is definitely the most advanced topic of the book and may take some time if you are not familiar with writing Go. We will also make sure only trusted plugins are used to ensure no malicious code gets used by mistake. We will cover the following in this chapter:

- Basics of Go

- Sample plugin source

- Building and testing your plugin

- Protecting yourself from bad plugins

By the end of this chapter, you should have a good foundation to customize, debug, and troubleshoot Packer plugins. We will also use this opportunity to write a plugin I've long thought should be included in Packer's supported builders: `systemd-nspawn`. There are already a few implementations of this builder in the community but I will start from scratch for this chapter because the builder is simply a directory in the local storage. Whether you love or hate systemd, I think that `systemd-nspawn` is one of its best features. This is a special type of `chroot` container where systemd runs a nested environment within the host's filesystem, almost like a systemd container on a physically bootable storage base. This is similar to cloud chroot builders supported by Packer but can be applied in any Linux environment running systemd. The resulting image is a location on a disk that can be captured as an archive, run as a `systemd-nspawn` container, or even booted from a disk or network via PXE virtual or physical servers. This offers a lot of flexibility since the same provisioners you use to customize VMs in the cloud can be used to build a network boot environment locally for VMs or even physical servers. Terminology can be confusing since `systemd-nspawn` often uses the terms *machine* and *container* interchangeably. First, let's cover the basics of Golang, also known as Go.

Technical requirements

This chapter requires knowledge of the Go programming language. A development environment can be set up for free on any machine that supports Go. A free IDE can be used to simplify the process if wanted. Atom and Visual Studio Code are two common IDE options. Building and testing a plugin during development can be tricky so a debugger environment may help as well as covered in this chapter.

Basics of Go

If you have never coded in Go before, I'll try to sum it up for you without offending the Go experts. Go is an open source programming language from Google inspired by C and C++ but featuring memory protection and automatic management of memory for the stack and heap. As a coder, this means you don't need to worry much about how you use memory but the Go runtime may take some extra time to rearrange or garbage-collect memory. Unlike scripting, it also means your code must have no syntax errors and compile successfully to build a binary. This includes strong static types and complete syntax during compilation. The good news is that performance and stability for Go are very good compared to scripting, which must be parsed and type-checked every time you run it. Go's performance is occasionally affected by memory garbage collection performed by the runtime, which scripts must also do anyway.

Goroutines are a simple mechanism for concurrency via lightweight processes (usually implemented as threads), though they are also managed by the Go runtime instead of the OS as threads typically are. Packer plugins usually won't need to manage goroutines or background processes as plugin processes are generally handled by the core Packer process. This simplifies things a bit, as we won't need to worry about concurrency or race conditions. The core Packer binary uses goroutines to run each source in parallel so, in most cases, goroutines won't be necessary in a plugin. Builds will happen concurrently unless otherwise stated in the template.

Since Packer shifted to HCL2, there are some important things to consider when building Packer plugins. Plugins are written in Go and use RPC for communication with the Packer binary. This means that your plugin must implement certain interfaces that the plugin library will hand over to Packer via a UNIX domain socket. The end product for plugins is a static Go binary that Packer will search for and call based on the path it's stored in. If you come across a Packer plugin and want to check what features it implements, you can check this with a basic command. The `describe` parameter will show you what the plugin supports in detail:

```
./packer-plugin-nspawn describe | jq
{
  "version": "0.1.0-dev",
```

```
    "sdk_version": "0.3.1",
    "api_version": "x5.0",
    "builders": ["machine"],
    "post_processors": [],
    "provisioners": [],
    "datasources": ["images"]
}
```

If you build your plugin and it doesn't show this, or it shows an error, then something is wrong at a high level. To help guide you on how to write plugins, there is a handy template provided by the Packer team to get you started. This plugin provides a simple starting example of what interface plugins need to implement. These templates are easily modified to accomplish what you need in a plugin, as long as you follow a few guidelines.

Since Packer added HCL2 support in addition to JSON support, the plugin needs to be able to handle both options. Go reflection and serialization do not natively support HCL2 so it's important that your plugin shows how to convert structures between JSON and HCL2 schemas. A lot of this is automated via a helper tool, which must be run manually to generate schema code. If you make a change to your plugin code and it crashes the build, then make sure to check your HCL2 mapping code has been updated. Go is statically typed and memory safe but it will crash if there is a segmentation fault or a dereference of unsafe memory. Such crashes can (and should) bring down the entire Packer process.

Pay very close attention to the comment in the scaffold plugin code. This especially applies to the file called data.hcl2spec.go. These files are auto-generated by a helper app that reads your code and produces a corresponding HCL2 schema in Go to match your code. If this isn't regenerated every time you make a code change, you may encounter issues. This is not a standard Go practice but only relates to HCL2. Notice the top line in data.hcl2spec.go:

```
// Code generated by "packer-sdc mapstructure-to-hcl2";
// DO NOT EDIT.
```

You will need to download this tool using Go and run it within your plugin directory. To install it, the scaffolding repo has a helpful target. Just run make install-packer-sdc or run the install command manually:

```
go install github.com/hashicorp/packer-plugin-sdk\
/cmd/packer-sdc@latest
```

This command may look a bit unconventional because it looks for comments at the top of your Go file to guide it. The first line of your plugin code (which defines the struct and schema for each plugin) should tell it how to run, specifying which type contains your config structure. When the packer-sdc command runs, it will parse the Go source file for the type specified and generate the corresponding HCL2 support code in the {PLUGIN}.hcl2spec.go file:

```
//go:generate packer-sdc mapstructure-to-hcl2 -type Config
```

It would be wise to add this command to your CI automation so that it happens automatically and you won't need to worry about running it manually for every code change you make. GitHub and GitLab support pre-commit or pre-receive hook events that can run this command before a commit is saved in version control. Any error in such a hook can block the accidental commit of broken code. See the documentation on server hooks for more details: `https://docs.gitlab.com/ee/administration/server_hooks.html`.

Packer plugins just need to implement a set of interfaces provided by the SDK. Then, RPC is used by Packer to communicate with your plugin binary, which should be stored in your `~/.packer.d/plugins/` directory. I find it simplest to drop a symlink to your binary as you develop:

```
ln -s ~/code/packer-plugin-nspawn/packer-plugin-nspawn \ ~/.packer.d/
plugins/
```

Each time you rebuild your plugin, it will automatically be picked up by the next Packer run. If you set `PACKER_LOG=1`, the trace output shows details about loading your plugin and querying what features it supports. It also shows where your RPC socket is created:

```
Received unix RPC address for /usr/bin/packer: addr is \ /tmp/packer-
plugin370850083
```

This socket is used to communicate with your plugin binary during a build, so the user will need permission to the path used. Also, if your plugin experiences a significant error or segmentation fault, it can crash the main Packer process, so be careful. In my experience, failing to rebuild the HCL2 schema for a plugin can cause a parse error or segmentation fault and even crash the main Packer process with a notice to file a bug against Packer. Hopefully, Packer will get better plugin isolation to prevent this later.

Go supports interop with standard C/C++ libraries via a feature called **CGo**, which allows dynamic linking of your code. We will skip this feature and use external commands to execute our plugin tasks. This keeps our plugin simple as it doesn't require external shared object libraries, but we will require an external command to be installed and available in the Packer environment. Now we've given an introduction to Go and how the plugin architecture works, let's delve into the sample code. We will build a single plugin provider to support a data source and a builder.

Sample plugin source

Always start with a current copy of the scaffold plugin repo. This has a few important files in the root directory, including `main.go`, `GNUMakefile`, and `go.mod`, which pins certain versions of the dependencies used by your Go module. It also declares a Go module, which you should rename to your module. Here, we use `packer-plugin-nspawn` for our module name. Have a look at the repository here: `https://github.com/jboero/packer-plugin-nspawn`.

The repo contains sample folders for each plugin type:

- `builder`: We will use this to start a `systemd-nspawn` machine
- `datasource`: We will use this to look up the machines and images available
- `post-processor`: We won't need one of these so we delete the directory
- `provisioner`: We also won't need one of these so we will delete it

The plugin registers itself in the very simple `main.go` file. Post-processors and provisioners may be added for custom actions but we won't need anything. Packer already provides common provisioners at its core. More likely, we would want to specify support for a custom communicator such as SSH so that our plugin can make use of Packer's existing post-processors and provisioners. The Packer plugin scaffold does not currently include a communicator, which would be a helpful inclusion. Otherwise, you may view an existing plugin repository to browse examples of working communicator code. A good example is the QEMU plugin: `https://github.com/hashicorp/packer-plugin-qemu/`.

It's super simple to register your implementation of the plugin interfaces and run the RPC process. Parameters to run your binary are even handled for you. This is how to tell Packer we support a builder called `machine` and a data source called `images` via our `nspawn` plugin:

```
func main() {
  pps := plugin.NewSet()
  pps.RegisterBuilder("machine", new(nspawn.Builder))
  pps.RegisterDatasource("images", \ new(nspawnData.Datasource))
  pps.SetVersion(nspawnVersion.PluginVersion)
  err := pps.Run()
  if err != nil {
    fmt.Fprintln(os.Stderr, err.Error())
    os.Exit(1)
  }
}
```

Note that directory structure and package naming are important. Packer will reference our plugin by `provider-plugin` naming standards. So, even though we register our data source with the name `images`, that data source is in the `nspawn` package. This means it must be referred to in Packer templates as `nspawn-images`. This can be confusing as Packer mixes the usage of this in logs.

Currently, `systemd-nspawn` doesn't seem to have a standard Go wrapper, but we'll keep this plugin nice and simple and just use `exec.Command` to control the standard tooling of the `machinectl` and `systemd-nspawn` commands. You can easily substitute any command you want to run as part of a plugin and customize the configuration structure for whatever you need, but it would be wise to use a Go binding when possible. These will need to be installed in the environment or else the plugin will fail with a **Command not found** message, which should be informative enough for the user.

Start with our simple `images` data source, where we list the available machines and images on the current host using the `machinectl` command. Explore the contents of the `datasource` folder. The minimal files you need are `data.go` and `data.hcl2spec.go`. The rest is for testing, and even `data.hcl2spec.go` is generated based on your `data.go` file:

```
$ tree datasource/
datasource/
└── nspawn
    ├── data_acc_test.go
    ├── data.go
    ├── data.hcl2spec.go
    └── test-fixtures
        └── template.pkr.hcl
```

All we really need to do is to get the JSON output from the `machinectl list-images` and `machinectl list` commands. We can do that simply with this statement. Luckily, the `machinectl` command supports a JSON output flag:

```
images, err := exec.Command( \
  "machinectl", "list-images", "-o", "json" \ ).CombinedOutput()
```

This simply gives a JSON list of the current images stored by default at `/var/lib/machines`. Note that, in this example, I have a Fedora Rawhide environment bootstrapped from Fedora's `dnf` package management tool:

```
sudo dnf groupinstall --installroot=/var/lib/machines/f38 \
    --releasever=38 --setopt=install_weak_deps=False \
    "Fedora Server Edition"
```

You may also pull a raw image file using the `machinectl` command:

```
$ machinectl pull-raw --verify=no \
https://download.fedoraproject.org/pub/fedora/linux/
releases/38/Cloud/x86_64/images/Fedora-Cloud-Base-38-
1.6.x86_64.raw.xz \
Fedora-Cloud-Base-38-1.6.x86-64

$ machinectl list-images -o json
[{"name":"rawhide","type":"directory","ro":false, \
  "usage":null,"created":1674929469692103,"modified":0}]
```

This is just an example, but we'll show you how to set up an initial image in the next section when we discuss testing your plugin. The default scaffold data source comes with a mock configuration option that does nothing. We have no need for configuration as this will all occur on the current host, but it may be a good idea to put a host option so you can let the plugin connect to other machines and access their `systemd-nspawn` images also.

What about the builder? In this case, we need at least one configuration option. What image should we deploy or what machine should we start for the environment? What environment variables should be added to the environment when it's started? These are potential configuration items we will need to capture as part of our HCL2 schema. We will require an image name to start a machine for the build, so that will be a mandatory requirement of our schema, causing Packer to error if an image has not been specified.

Let's take a look at the `builder` directory:

```
$ tree builder/
builder/
└── nspawn
    ├── artifact.go
    ├── builder_acc_test.go
    ├── builder.go
    ├── builder.hcl2spec.go
    ├── communicator.go
    ├── step_say_config.go
    └── test-fixtures
        └── template.pkr.hcl
```

This is slightly more complex than the data source. We need to worry about build stages and artifacts. Also, `systemd-nspawn` tends to be very strict on permissions. Starting a machine requires root access, and even with root access, there is a known bug where SELinux must be disabled (or set to permissive). So, for this example, we will run Packer as root. If using automation, make sure the runner is on an isolated or dedicated machine. If you don't have access to root and can't set appropriate SELinux labels for `machinectl` to start and stop machines, this plugin may not work for you. We will let standard error handling report this to the user. When you start a `systemd-nspawn` environment with `machinectl`, it starts a complex process with a templatized systemd unit for `systemd-nspawn@.service`. For example, `machinectl start rawhide` actually does the following chain of operations:

1. First, it runs:

    ```
    machinectl start rawhide
    ```

2. This then runs:

    ```
    systemctl start systemd-snawpn@rawhide.service
    ```

3. This then runs:

    ```
    systemd-nspawn --quiet --keep-unit --boot \
        --link-journal=try-guest --network-veth -U \
        --settings=override --machine=rawhide
    ```

If a successful setup is done, this fully boots a `systemd-nspawn` environment with the specified image within our host. This typically takes about 2 seconds, which is much quicker than booting a full VM via systemd. Note the image is not implicitly cloned to a new one. We use the "prepare" step to clone this image and present a fresh environment to Packer for each build. The difference between a `systemd-nspawn` container (aka machine) and a Docker container is that systemd boots a full environment and behaves as `init` for the container. Multiple processes can run within the container, just like a normal host. The `systemd-nspawn` environment acts and behaves completely like a standalone systemd host, whereas traditional Docker and Podman containers usually just run one process and rely on the host's `init` to manage failures and hardware resources. If the image is configured to run SSH, then the standard SSH communicator may be used, but there's a better option available for `systemd-nspawn`, which is just a local folder. Normal builders can just rely on a common communicator such as SSH or WinRM to connect during provisioners. Remember, Docker has its own communicator because it will just connect to the Docker container locally when no SSH connection is required. To use provisioners for a `systemd-nspawn` builder, we would need to tell Packer how to connect to the local container. Luckily for us, this is pretty straightforward since it's just a local folder and container similar to Docker. The tricky part is that custom communicators are rarely needed for plugins so they aren't very well documented. The key is to implement the simple communicator interface. This offers the following options to be implemented:

- `start`: How to start a remote command
- `upload`: How to upload a file
- `uploadDir`: How to upload a directory
- `download`: How to download a file
- `downloadDir`: How to download a directory

The `upload` and `download` operations are super straightforward with the `systemd-nspawn` builder. All we need to do is copy contents to and from the container path at `/var/lib/machines/`, which is easy and, technically, can be the same code for a directory or a file. Uploading and downloading are done synchronously, such that execution doesn't resume until your implementation is finished. The `start` function is a bit trickier. This is meant to execute asynchronously in the background and it's up to the developer to handle the finished event and wait for the command to finish.

To learn about communicators, it's helpful to view the source code for existing plugins. The Docker plugin's communicator is a great case because it often makes use of `exec.Command` to run Docker commands for operations. Note the importance of telling Packer how to create temporary space for your communicators. Copy operations and file manipulations aren't performed directly. Packer allocates temporary space to stage data and scripts before applying it. Obviously, if a script contains template values, it needs to be rendered for a resulting script. Also, if a provisioner script is applied to multiple sources, then multiple versions of the script may be needed. By default, the temporary space consists of `tmpfs` at `/tmp` and uses randomized names to distinguish resources. If you need to allocate temporary space, you can override that location. If any one of these steps is not implemented correctly for your plugin, there is a chance your plugin won't respond correctly to an RPC call. This may result in the build failing with an unexpected EOF and should be a sign that something is missing from your code.

Building and testing your plugin

The GNUMakefile scaffolding also includes simple test support. It would be wise to modify any test Packer templates to demonstrate every feature of your plugin fully. There is a helpful test target to put the test templates and mock data through your plugin. When ready, you can simply run make test and you will get an informative report of failures or issues:

```
$ make test
?          packer-plugin-nspawn      [no test files]
ok         packer-plugin-nspawn/builder/nspawn       0.047s
ok         packer-plugin-nspawn/datasource/nspawn    0.036s
?          packer-plugin-nspawn/version      [no test files]
```

Having this added to automation would also be a very wise move. A pre-commit hook can prevent broken code from being committed and also log one of these tests for compliance.

To set up a systemd-nspawn image, you can simply use your OS packaging management. The examples used for development were built from Fedora's DNF packager, but Debian users have the option of the debootstrap command:

```
debootstrap --include=systemd stable /var/lib/machines/debian
```

This command sets up a basic Debian environment with systemd ready for an nspawn container. This does not contain a kernel, which would be required for actually booting or PXE booting an environment. Kernel and boot support would need to be added either to the root image or via your Packer template. Let's see how to do the same thing for Fedora or an Enterprise Linux-based distribution:

```
sudo dnf --installroot=/var/lib/machines/rawhide \
  --releasever=rawhide \
  --nogpgcheck install fedora-release systemd glibc
```

Testing with the sample Packer template gives us the following data source and builder example. This will just output the local images and the machines on the current host. Since there is no configuration necessary to list the local images and machines, we can skip the default mock configuration parameter because it is unused:

```
data "nspawn-images" "test" {}
```

A sample builder is as easy as specifying an image. This image must exist on the local host or an error will stop the build. There are three calls to `machinectl` during a build, which require privilege escalation if the Packer build is not run as root. Running Packer in a graphical session causes `machinectl` to prompt a dialog for your password all three times. It would be much better to run this as root or adjust the `sudoers` file to allow a Packer user elevated privileges for Packer without a password:

```
source "nspawn-machine" "basic-example" {
  image = "rawhide"
}
```

This test should successfully start and stop a `systemd-nspawn` container, but if there are no communicators available, then provisioners won't work. An artifact could also be created with an archive of the image, but for now, we just leave the build image in `/var/lib/machines/` where they are kept by default.

Protecting Packer from bad plugins

The release cycle is important with plugins. Developing a plugin locally is one thing but using it in production should require a signed release. You can tag and release a version of your repo for the community to use. You should always sign releases to ensure malicious code doesn't make its way into your plugin. GPG keys can be used to sign a release, and public keys can verify that the plugin matches what you shipped it with.

In addition to plugin signatures, it's important to make sure your Packer template pins the correct, or at least minimal, release for your plugin in case the functionality changes. Remember, this can be specified at the top of a template. Also, for machines that don't have the source code for building your plugin, it's important to configure the source path. This will require a `packer init` command before the first build you attempt so that Packer can download the right release into the `$HOME/.packer.d/plugins` directory:

```
packer {
  required_plugins {
    nspawn = {
      version = ">=v0.1.0"
      source  = "github.com/jboero/nspawn"
    }
  }
}
```

Community plugins are not supported by HashiCorp, and generally, the entire product is community supported, so be sure to report issues against a plugin's repository if you have issues. Most plugins are supportive and welcome pull requests if you find a fix or add a feature.

Summary

Packer plugins are an advanced topic, and here we did a super simple example written in Go. The current versions of Packer use RPC for communicating with the main Packer process. This means that plugins could technically be written in any language that supports RPC, but the great SDK complemented by the Packer team's scaffold plugins in Go makes a fairly complex concept simple via standard interfaces. It is possible to merge a popular plugin into the main Packer binary but this requires approval from the Packer engineering team. There is no reason a Packer plugin can't be used in production environments so long as it is packaged properly as a signed Git release and cached locally via `packer init`.

Data source plugins are quite simple, with inputs and outputs. Builders and provisioners are more complex plugins that may require implementing an additional communicator for standard provisioners to access your build environment. Make sure, when writing a production plugin, that you also write test templates that fully demonstrate all features in your plugin. This will help prevent the release of a buggy or invalid plugin.

The sample source code for this chapter is in a separate repository because of the required naming conventions for a plugin. We implemented a super basic builder and a partial communicator to demonstrate the basic mechanics of a communicator. To learn more about plugins, it is a good idea to check out the source code for existing plugins, which are all public and open source.

Grand conclusion

As we come to a close, I hope you've learned some helpful tricks to add to your Packer templates. Even though Packer is one HashiCorp's simplest tools and projects, it's incredible what you can do with it. Often, people think of Packer as just a tool for building cloud images. Packer is that and much more, with the ability to build and test everything from mobile applications on ARM devices to embedded microcontroller applications and containers. All of this can be done within a single template or template directory. Remember, a lot of old Packer code is written in a legacy JSON format that is still supported for backward compatibility, but new templates should be written in HCL2 or the new `pkr.json` format.

Don't forget the cost savings from making sure your application is built and tested for alternative architectures such as ARM or AArch64 and AWS Graviton. This step alone can save you 50% or more on your compute spend. Also, being built across multiple cloud providers keeps your application agile when outages or cost-reduction opportunities come up. There is no reason you can't make a simple addition to your existing templates and try this today.

Thanks to Packer's QEMU builder, you don't even need to have a cloud account to test any of these environments locally. This includes running any of the example code in this chapter on a low-cost SBC such as Raspberry Pi or other student devices. QEMU is a great open source project that levels the architecture playing field and allows a large number of architectures to be emulated from any one of those architectures.

Best practices with a GitOps workflow and automation make easy work of developing Packer templates with team collaboration. In fact, setting up automation would be a wise first step when working with Packer. Setting up HCP Packer for metadata should also be an early strategy to track your image evolution and patch lifecycle.

Packer images are usually deployed with Terraform to simplify lift-and-shift migrations. Packer has many layered features, including provisioners, inputs, and outputs. Packer enables *Images as Code* for best practices in VM deployment locally or in the cloud. Terraform will deploy your resources much quicker when the images are pre-built and tested. The next step of your HashiCorp journey generally involves learning Terraform.

Index

V

packtpub.com

Subscribe to our online digital library for full access to over 7,000 books and videos, as well as industry leading tools to help you plan your personal development and advance your career. For more information, please visit our website.

Why subscribe?

- Spend less time learning and more time coding with practical eBooks and Videos from over 4,000 industry professionals

- Improve your learning with Skill Plans built especially for you

- Get a free eBook or video every month

- Fully searchable for easy access to vital information

- Copy and paste, print, and bookmark content

Did you know that Packt offers eBook versions of every book published, with PDF and ePub files available? You can upgrade to the eBook version at packtpub.com and as a print book customer, you are entitled to a discount on the eBook copy. Get in touch with us at customercare@packtpub.com for more details.

At www.packtpub.com, you can also read a collection of free technical articles, sign up for a range of free newsletters, and receive exclusive discounts and offers on Packt books and eBooks.

Other Books You May Enjoy

If you enjoyed this book, you may be interested in these other books by Packt:

Learning DevOps - Second Edition

Mikael Krief

ISBN: 978-1-80181-896-4

- Understand the basics of infrastructure as code patterns and practices
- Get an overview of Git command and Git flow
- Install and write Packer, Terraform, and Ansible code for provisioning and configuring cloud infrastructure based on Azure examples
- Use Vagrant to create a local development environment
- Containerize applications with Docker and Kubernetes
- Apply DevSecOps for testing compliance and securing DevOps infrastructure
- Build DevOps CI/CD pipelines with Jenkins, Azure Pipelines, and GitLab CI
- Explore blue-green deployment and DevOps practices for open sources projects

HashiCorp Infrastructure Automation Certification Guide

Ravi Mishra

ISBN: 978-1-80056-597-5

- Effectively maintain the life cycle of your infrastructure using Terraform 1.0
- Reuse Terraform code to provision any cloud infrastructure
- Write Terraform modules on multiple cloud providers
- Use Terraform workflows with the Azure DevOps pipeline
- Write Terraform configuration files for AWS, Azure, and Google Cloud
- Discover ways to securely store Terraform state files
- Understand Policy as Code using Terraform Sentinel
- Gain an overview of Terraform Cloud and Terraform Enterprise

Packt is searching for authors like you

If you're interested in becoming an author for Packt, please visit authors.packtpub.com and apply today. We have worked with thousands of developers and tech professionals, just like you, to help them share their insight with the global tech community. You can make a general application, apply for a specific hot topic that we are recruiting an author for, or submit your own idea.

Share Your Thoughts

Now you've finished *HashiCorp Packer in Production*, we'd love to hear your thoughts! Scan the QR code below to go straight to the Amazon review page for this book and share your feedback or leave a review on the site that you purchased it from.

https://packt.link/r/1803246855

Your review is important to us and the tech community and will help us make sure we're delivering excellent quality content.

Download a free PDF copy of this book

Thanks for purchasing this book!

Do you like to read on the go but are unable to carry your print books everywhere? Is your eBook purchase not compatible with the device of your choice?

Don't worry, now with every Packt book you get a DRM-free PDF version of that book at no cost.

Read anywhere, any place, on any device. Search, copy, and paste code from your favorite technical books directly into your application.

The perks don't stop there, you can get exclusive access to discounts, newsletters, and great free content in your inbox daily

Follow these simple steps to get the benefits:

1. Scan the QR code or visit the link below

https://packt.link/free-ebook/9781803246857

2. Submit your proof of purchase
3. That's it! We'll send your free PDF and other benefits to your email directly